SPECIAL PAPERS IN PALAEONTOLOGY NO. 79

NAUTILOIDS BEFORE AND DURING THE ORIGIN OF AMMONOIDS IN A SILURO-DEVONIAN SECTION IN THE TAFILALT, ANTI-ATLAS, MOROCCO

BY

BJÖRN KRÖGER

with 16 plates and 23 text-figures

THE PALAEONTOLOGICAL ASSOCIATION

LONDON

March 2008

CONTENTS

[Special Papers in Palaeontology, 79, 2008, pp. 1–110]

Abstract: The non-ammonoid cephalopod fauna of the Siluro-Devonian section of Filon Douze, in the southern Tafilalt, Morocco is described. The section spans a sedimentary succession of predominantly argillites with intercalated cephalopod limestones of Ludlow–Eifelian age with a thickness of *c.* 450 m. More than 2000 cephalopods were collected bed by bed and comprise 52 genera (17 new) and 86 species (39 new). Only one discosorid occurs, the new taxon *Pseudendoplectoceras lahcani*. The oncocerids are highly endemic, since out of nine recorded genera five are new: *Cerovoceras, Mutoblakeoceras, Orthorizoceras, Tafilaltoceras* and *Ventrobalashovia*. Only one of 13 oncocerid species is known from elsewhere and nine are new: *Bohemojovellania adrae, B. obliquum, Brevicoceras magnum, Cerovoceras brevidomus, C. fatimi, Jovellania cheirae, Mutoblakeoceras inconstans, Tafilaltoceras adgoi* and *Ventrobalashovia zhuravlevae*. Three actinocerids occur, two of which, *Metarmenoceras fatimae* and *Deiroceras hollardi*, are new. Within the Pseudorthoceratida the new taxa *Cancellspyroceras, Geidoloceras ouaoufilalense, Subdoloceras atrouzense, S. tafilaltense,* and *S. engeseri, Subormoceras erfoudense* and *S. rissaniense* are erected. The pseudorthoceratid species *Neocycloceras termierorum, Spyroceras cyrtopatronus, S. latepatronus* and *Sulcoceras longipulchrum* are also erected. New genera and species of the Orthocerida are *Adiagoceras taouzense, Angeisonoceras reteornatum, Chebbioceras erfoudense, Infundibuloceras brevimira, I. longicameratum, I. mohamadi, Pseudospyroceras reticulatum, Theoceras felondouzense* and *Tibichoanoceras tibichoanum*. Additionally, the orthoceratid species *Hemicosmorthoceras aichae, Orthocycloceras tafilaltense, Plagiostomoceras lategruenwaldti, P. reticulatum, Sichuanoceras zizense* and *Temperoceras aequinudum*, and the bactritoid species *Devonobactrites emsiense* are erected, and 22 species are transferred to different genera. The stratigraphical section is described in detail and the depth of deposition of key horizons is estimated. Cephalopods occur mainly in three different facies types: (1) massive limestones, silty shales and marls with a bivalve-orthocerid association, reflecting (par-)autochthonous conditions in a distal environment below storm wave base; (2) proximal tempestites with a bivalve-orthocerid association, reflecting (par-)autochthonous conditions in a distal environment above storm wave base; and (3) marls and nodular limestones containing orthocones and a diverse benthos reflecting a well-oxygenated environment below storm wave base. The bivalve-orthocerid association is exclusively pre-Pragian. The bivalve-orthocerid storm beds abruptly disappear at the top of the Lochkovian. The Pragian, Emsian and Eifelian sediments invariably contain cephalopods together with a highly diverse benthos. A significant increase in cephalopod richness and taxonomic distinctness occurs in the uppermost Lochkovian tempestites. The Lochkovian/Pragian boundary also marks a profound change in the morphological composition of the cephalopod association. In the uppermost Lochkovian several cephalopod taxa, which were adapted to the low energy needs that dominated during the late Silurian and earliest Devonian, have their last occurrence. In post-Lochkovian strata cephalopod morphotypes that were adapted to energy-intensive buoyancy regulation dominate. Finally, in late Pragian and Zlíchovian deposits, bactritoids *sensu stricto*, ammonoids, coiled nautiloids and several pseudorthoceratids have their first occurrence. These groups dominated in the late Palaeozoic. Therefore, the changes at the Lochkovian/Pragian boundary resulting in better conditions for life on the seafloor can be interpreted as a precondition that led to the landmark evolutionary innovations in the Zlíchovian.

Key words: Cephalopoda, bactritoid origin, cephalopod limestones, Silurian, Devonian, Morocco.

IN the early Middle Devonian (Zlíchovian) the ammonoids appeared and quickly evolved to be the dominant shelled cephalopods until the end of the Cretaceous Period. The ammonoid origin and initial radiation, including its timing and morphological pathways, are comparably well known (Erben 1960, 1964; Bogoslovsky 1969; Korn and Klug 2003). However, the ammonoids did not enter an empty cephalopod universe. The seas of the Early Devonian were inhabited by a multitude of cephalopods prior to their appearance and the incipient ammonoid faunas had to share their ecosystem with several close relatives. Bactritoids and Orthocerida are the closest relatives

BJÖRN KRÖGER

Université Lille 1, Laboratoire Géosystèmes (UMR 8157 CNRS), UFR des Sciences de la Terre, bâtiment SN5, 59655 Villeneuve d'Ascq Cedex, France; formerly Museum für naturkunde der Humboldt Universität zu Berlin, Invalidenstraße 43, D-10115 Berlin, Germany; e-mail: bjoekroe@gmx.de

 doi: 10.1111/j.1475-4983.2008.00764.x

of ammonoids, since they share a small, spherical, embryonic shell and a very narrow siphuncle that lacks deposits. Finds of orthocerid radulae (Mehl 1984; Gabbott 1999) suggest not only similar feeding habits but also close phylogenetic relationships between ammonoids and orthocerids.

The evolution of these close relatives just before, during, and after the appearance of ammonoids has never been investigated. A study of the literature reveals that no information is available on the diversity and commonality of bactritoids in the Late Silurian–Middle Devonian. It is not even clear when the direct ancestors of the ammonoids appeared. Ristedt (1981) described a single species of a Ludlow bactritoid from Bohemia, but questioned a direct ancestry of the Devonian bactritoids from that Silurian specimen. Beside this species, no unquestionable pre-Zlíchovian bactritoids have been reported. A similar situation exists in knowledge of the evolution of post-Lochkovian orthocerids. Orthocerids are the dominant elements in the cephalopod limestone biofacies (Kříž and Ferretti 1995; Kříž 1998) of peri-Gondwanan low latitudes. The orthocerids of this facies have been intensively investigated (Babin 1966; Ristedt 1968; Serpagli and Gnoli 1977; Gnoli 1982; Holland *et al.* 1994; Histon 1999; see also references in Gnoli 2003). However, these orthocerid-dominated limestones disappear towards the top of the Lochkovian (Kříž 1998, 1999; Gnoli 2003) and the orthocerid-bearing limestones of the post-Lochkovian have rarely been investigated with a focus on orthocerids. Although post-Lochkovian Devonian orthocerids have been described in several papers (e.g. Foerste 1926, 1927, 1929; Flower 1936, 1939, 1949; Balashov and Kiselev 1968; Zhuravleva 1978; Gnoli 1982), the stratigraphical control of the occurrences of the cephalopods described is often poor, making detailed comparisons with ammonoid occurrences difficult. Therefore, from the study of the literature it is impossible to evaluate how non-ammonoid cephalopods, specifically the orthocerids, developed during the initial ammonoid radiation in the earliest Emsian. The current work is intended to contribute to the knowledge of stratigraphical occurrences of non-ammonoid cephalopods within this critical time interval.

During two field trips in 2003 and 2005, I collected more than 2000 cephalopod specimens, bed by bed, from a section in the Tafilalt of south-east Morocco spanning the Ludfordian–early Eifelian. For the first time, non-ammonoid cephalopod occurrences were recorded against a precisely controlled stratigraphical record in post-Lochkovian Devonian sediments. Although the cephalopods were collected only at a single locality, it is assumed that the occurrences represent some general pattern. I draw this conclusion from the widespread geographical occurrence of orthocerid beds throughout Morocco and North Africa and of the commonality of many faunal elements across the Rheic Ocean from Bohemia to the Carnic Alps, Armorica, Sardinia and Morocco (Gnoli 2003). Specific orthocerid beds can be traced in North Africa from Mali to Algeria (Kříž 1998; Gnoli 2003). This detailed report of the stratigraphical occurrences at this single locality is intended to serve as a reference for future investigations in other regions of the world.

TERMINOLOGY

Open nomenclature. The open nomenclature follows the recommendations of Bengtson (1988). Thus, '?' following a generic or specific name denotes uncertainty of the determination because of poor preservation or poor material; quotation marks '…' indicate that the genus or species is regarded as obsolete in the current context; 'cf.' indicates that the specific identification is provisional and further material is needed in order to establish a new species or to confirm the present assignment.

Higher taxa. I use a variety of endings to distinguish related higher taxa: nautilids, nautilidans and nautiloidans; and bactritids, bactritidans and bactrioids. The endings '-ids' indicate the specific family, '-idans' a subgroup of the specific order that does not coincide with a particular family, and '-oids' a group that in some aspects is similar to the specific family.

Morphological terms. The conch is divided into a left and a right side from a perspective viewed in the direction of growth along the dorsoventral axis. The cross-section of the conch is *compressed* when it is higher than wide, and *depressed*, when wider than high. The shell surface is ornamented by *striae* when two successive grooves form a band in between; by *lirae* when raised lines occur; by *ridges* when the conch is locally thickened forming acute or rounded waves (Text-fig. 1).

The shape of the septal neck is defined by the angle of inflection relative to the direction of the septal surface, which is usually approximately perpendicular to the growth axis. The septal necks are *achoanitic* when very short or lacking and too short to detect the angle of inflection; *loxochoanitic* when the tips of the necks point at an angle of less than 90 degrees; *orthochoanitic* when the angle is nearly 90 degrees; *suborthochoanitic* when it is 90–180 degrees; and *cyrtochoanitic* when it is more than 180 degrees. The septal necks are *recumbent* when the brim touches the adapical surface of the related septum.

The siphuncle is *subcentral* when it is removed from the growth axis by a distance less than its own diameter,

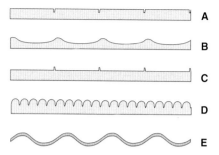

TEXT-FIG. 1. Illustrations of types of surface ornamentation in median section. A, striae. B–C, lirae. D, ridges. E, undulations. Not to scale.

eccentric when it is removed from the growth axis by a distance more than its own diameter, and *marginal*, when it touches the shell wall. The cameral deposits are *hyposeptal* when covering the adapical septal surfaces; *episeptal*, when covering the adoral septal surfaces; and *epichoanitic* when covering the septal necks (Text-fig. 2). The endosiphuncular deposits are *annulate* when arranged symmetrically, or nearly so, around the septal perforation, and *parietal* when annuli are either adorally or adapically elongated.

DIVERSITY MEASURES

The number of specimens collected from each bed differs greatly (see Appendix), since from some beds several hundreds of specimens were recovered (the maximum was more than 550 specimens from bed EF), whereas others yielded only a few specimens (only five from the

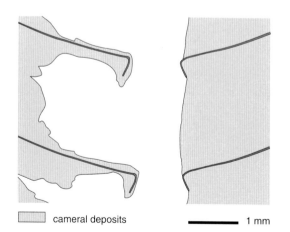

cameral deposits ▬▬▬ 1 mm

TEXT-FIG. 2. Epichoanitic deposits, camera lucida drawing of median section of *Arthrophyllum vermiculare*; MB.C. 9648; Pragian, Filon Douze section, Tafilalt, Morocco. Note the coverage of entire septal necks by cameral deposits. Ventral is to the right.

Deiroceras limestones). The number of specimens collected reflects their accessibility rather than their real abundance. Collecting in clay and in siltstones, where the fossils washed out and accumulated on the bed surface, was easier than collecting from limestones that in some cases only cropped out with surfaces perpendicular to the bedding planes. Moreover, I did not collect every single specimen in limestones but only when easily accessible and with a focus on reaching the greatest richness during collecting. In contrast, I collected every specimen in the clay, siltstones, and marls regardless of their preservation and type. The diversity estimation is further complicated by different preservation in specific beds. The fauna of bed OP-M is recorded only by internal moulds of the conchs, that of bed EF by limonitic steinkerns, and other beds preserved a fauna with more or less well-preserved but recrystallized conchs. These differences in preservation and collecting practice were minimized by sample standardization, and by excluding some beds from diversity estimation. Only beds OP-M, PK, OJ-I, KMO-I–KMO-IV and EF were considered for diversity estimation. I collected a large number of specimens from these beds; every fragment, regardless of preservation or type.

Species richness was calculated using the method of Chao (1984) following Chazdon *et al.* (1998) (Table 1). The advantage of this method is that it is not based on the parameters of a species abundance model that has previously been fitted to the data. Magurran (2004) demonstrated that the Chao estimators are among the most robust diversity measures developed in recent years. The absolute number of species in a sample is estimated as

$$S_{chao1} = S_{obs} + (F_1^2 / 2F_2)$$

where S_{obs} is the number of species in the sample, F_1 is the number of observed species represented by a single specimen, and F_2 is the number of species represented by two specimens. For the calculation of S_{chao1} I used EstimateS software (Colwell 2005). Nevertheless, the richness estimate is dependent on sampling intensity; it is generally underestimated in samples unless a very large sampling size is used. Magurran (2004) proposed a simple check for the accuracy of the estimations. When the total sample is subdivided into two subsamples at random and an estimate of the richness of the subsamples is obtained separately, the results should be consistent with those obtained from the total sample. I checked this for the samples of the Filon Douze section (Text-fig. 3; Table 2). The results show that a sufficient sample size is reached only in beds OP-M, PK and EF, the remaining samples consistently under-evaluate the richness of the total sample. However, Text-figure 3 also shows that even in the case under-estimating total richness, the general pattern of the diversity trend is not altered.

TABLE 1. Diversity measures of selected beds in the Filon Douze section.

	OP	PK	OJ-I	KMO-I	KMO-II	KMO-III	EF
Occurrences (total)	441	103	302	84	129	40	560
S_{obs} (Mao Tao)	13	3	17	10	13	13	7
S_{obs} 95% confidence lower b.	6	0	10	5	6	6	2
S_{obs} 95% confidence upper b.	20	6	24	16	20	51	12
$S_{Chao\ 1}$	21	4	18	18	17	25	7
$S_{Chao\ 1}$ 95% confidence lower b.	14	3	17	11	14	15	7
$S_{Chao\ 1}$ 95% confidence upper b.	78	11	27	74	42	81	7
Shannon evenness	1.26	0.15	1.33	1.55	1.71	2.03	0.57
Simpson evenness	2.55	1.06	2.45	3.49	3.8	5.77	1.32

Abbreviations: S_{obs} (Mao Tao), number of species expected given the total occurrences after Colwell *et al.* (2004); S_{obs} 95% confidence lower boundary/upper boundary, lower and upper boundary of 95% confidence level of S_{obs} (Mao Tao); $S_{Chao\ 1}$, Chao 1 richness estimator after Chao (1984); $S_{Chao\ 1}$ 95% confidence lower/ upper boundary, lower and upper boundary of 95% confidence level after Chao (1987). All values are calculated with EstimateS (Colwell 2005).

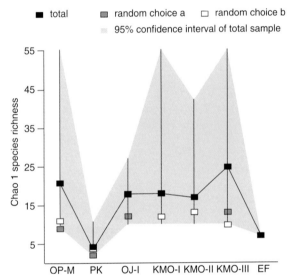

TEXT-FIG. 3. Chao 1 richness of total samples and two random subsamples. Bars indicate 95% confidence levels of the total sample. Calculated richness in random subsamples is considerably lower than in total sample from bed KMO-I–III, but the general diversity pattern with low values in beds PK and EF is preserved; for data, see Tables 2–3.

In addition to the Chao richness estimator, I calculated the taxonomic distinctness following Clarke and Warwick (1998, 1999) (for data, see Table 3). Taxonomic distinctness provides a measure of the relatedness of the species in a sample. It can be interpreted as an approximation of disparity in cases where few or no homoeomorphic taxa occur, which is the case in the Filon Douze section. The taxonomic distinctness index Δ^+ is calculated as:

$$\Delta^+ = (\Sigma\Sigma_{i<j}\omega_{ij})/(s(s^{-1})/2)$$

where s is the number of species in the sample and ω is the taxonomic path length between the species i and j. I weighted the path length assigned to each level of the taxonomic hierarchy as 1. Thus, each step up in the hierarchy of a shared taxonomic level (from species to genus, family and order) increments have the value 1.

Additionally I calculated the Shannon index H' and the Simpson index D as evenness measures using EstimateS software (Colwell 2005) in order to evaluate the heterogeneity of the samples.

THE LOCALITY

The succession investigated is a natural outcrop across a dry valley in the Tafilalt, Anti-Atlas, a region in southeastern Morocco south of the city of Errachidia, between Erfoud and the military post Taouz (Text-fig. 4). The section is located *c.* 10 km north-west of Taouz, spanning the entire unnamed valley south of the ridges of the Jebel Ouaoufilal at the section 30°56.592/004°03.059 – 30°57.095/004°02.290 slightly south of the abandoned lead mine of Filon Douze. The section of Hollard (1977, fig. 4) represents the same locality, but was probably measured several hundred metres to the north of Filon Douze. The sections at Jebel Ouaoufilal of Klug (2001, 2002), and Klug *et al.* (2008) are nearly identical to the Early and Middle Devonian intervals of the section considered herein.

SEDIMENTARY SUCCESSION AND STRATIGRAPHY

The interval investigated comprises a sedimentary succession of *c.* 450 m spanning beds from Ludfordian 'Orthoceras' limestones to early Eifelian *Pinacites* Limestone (Bultynck and Walliser 2000). The Late Silurian–Lochko-

TABLE 2. Diversity measures of two random subsamples of the total occurrences in the Filon Douze section given in Table 1.

	OP	PK	OJ-I	KMO-I	KMO-II	KMO-III	EF
Random Sample A							
Occurrences (total)	206	62	79	68	36	17	31
Chao 1	9	2	12	18	13	13	7
Chao 1 95% confidence lower b.	9	2	10	10	10	10	7
Chao 1 95% confidence upper b.	13	2	29	60	38	35	7
Random Sample B							
Occurrences	147	41	220	29	83	21	251
Chao 1	11	4	18	12	13	10	7
Chao 1 95% confidence lower b.	5	3	15	9	11	8	7
Chao 1 95% confidence upper b.	17	16	40	28	25	27	7

Note the lower calculated richness values in all beds, exclusive of EF. However, the general pattern with low richness values in bed PK and EF is preserved in the random samples. Abbreviations as in Table 1.

TABLE 3. Taxonomic distinctness of selected beds in the Filon Douze section calculated after Clarke and Warwick (1998, 1999).

Bed	Taxonomic distinctness ($^+$)
OP	2.7
OSP-I+II	2.6
PK	2.7
OJ-I+II	3.4
KMO-I	3.7
KMO-II	3.5
KMO-III	3.4
EF	3.3
KMO-IV	3.8
Erbenoceras Limestone	3.9
jugleri Limestone	3.5

Taxonomic distinctness provides a measure of the relatedness of the taxa in a sample. Note the increased values in beds of the *Jovellania* Limestones and above.

vian beds of the southern Tafilalt are comparatively poorly known: only short descriptions were provided by Hollard (1977). A detailed stratigraphical analysis of the early Emsian–middle Eifelian succession of sections of the southern Tafilalt was provided by Klug (2001, 2002). Parts of the Lochkovian succession of the northern Tafilalt were described by Alberti (1969), Belka *et al.* (1999) and Bultynck and Walliser (2000). Several key horizons can be distinguished at the Filon Douze section.

The sedimentary succession is divided here into five portions, indicated from oldest to youngest by 1–5 below.

1. Temperoceras limestone succession (Text-fig. 5)

Biostratigraphy. Polygnathoides siluricus conodont Biozone, early Ludfordian after Hollard (1977) and Brachert *et al.* (1992); but *Ozarkodina snajdri-Ozarkodina crispa* conodont Biozone, late Ludfordian after Stefan Lubeseder

(pers. comm. 2006). The unit was termed '*Orthoceras* Limestone' by Hollard (1977, p. 182).

Thickness. Approximately 5 m.

Description. The base of the unit consists of a massive 2-m-thick, dark, biodetrital pelagic mudstone (bed OP). At the base of the bed large load casts occur. There are several horizons with *Temperoceras*, which are *c.* 20–50 mm in cross-section. The orthocones are predominantly orientated in an east/west direction (80/260 degrees) and the apices are directed towards the west.

Above OP a succession of dark mudstones to wackestones and thin argillites with a thinning-up tendency in the limestones occurs. Frequently orthocones occur which are smaller than in the underlying massive limestone (0.1–0.3 m in cross-section); the bivalve *Cardiola* is common. Towards the top of the bed the orthocones and bivalves occur less frequently, and finally are almost absent.

The argillaceous limestone changes to a silty claystone (bed OP-M) towards the top. This bed is about 1 m thick and contains masses of small orthocones (cross-section *c.* 10 mm), predominantly *Arionoceras* and *Kopaninoceras*. Within the claystone large dark concretions (floatstone) occur (with horizontal diameters up to 0.5 m and a thickness up to 0.3 m); these contain a different fauna of orthocones with predominantly *Kopaninoceras* and *Plagiostomoceras*. The orthocones in the concretions are poorly sorted and small fragments (cross-section <2 mm) occur beneath large fragments (cross-section >100 mm). Remarkably, the large orthocones are often vertically orientated.

Above the marl a reddish–brownish, 0.5-m-thick, sandstone occurs that possesses scour and fill structures and an erosional surface at the top. Occasional small, indeterminable orthocones occur in this bed.

Above the sandstone a thin (0.02–0.05 m) biodetrital limestone occurs that contains masses of pelmatozoan

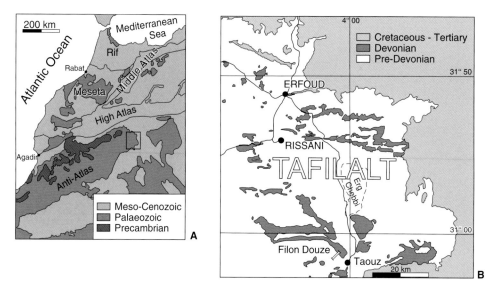

TEXT-FIG. 4. A, geological map of north-east Africa; rectangle shows area of B, which is a geological map of the Tafilalt, Anti-Atlas, Morocco, and shows the location of the Filon Douze section in the south.

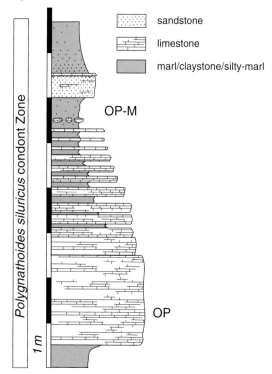

TEXT-FIG. 5. Section of the *Temperoceras* limestone succession, Ludfordian (Ludlow), Filon Douze, Tafilalt. This section is rich in orthoceridans (mainly *Arionoceras*, *Kopaninoceras* and *Temperoceras*) and bivalves.

debris, bivalves and small orthocones (mainly *Arionoceras*, cross-section <10 mm); it is terminated by a limonite-impregnated, irregular omission surface.

The limestone is covered by a sandy siltstone more than 100 m thick whose arenitic content decreases quickly within the basal few metres.

2. Scyphocrinites *limestone succession* (Text-fig. 6)

Biostratigraphy. Spathagnathodus steinhornensis/Ligonodina detorta conodont Biozone–*Icriodus postwoschmidti/woschmidti* conodont Biozone, latest Přídolian–earliest? Lochkovian after Haude and Walliser (1998). Because the *postwoschmidti/woschmidti* conodont Biozone spans an interval from latest Silurian to earliest Devonian and because no new graptolite data are available, the precise position of the Siluro-Devonian boundary is not known at Filon Douze. Hollard (1977) reported *Monograptus* ex gr. *lochkovensis* from his bed TM 714, which is considered similar to bed S-II of this investigation. This graptolite is indicative of the *Ozarkodina remscheidensis* interval–*L. detorta* conodont Biozone of the latest Silurian. Belka *et al.* (1999) reported a conodont fauna within the *Scyphocrinites* limestone succession from eastern Tafilalt that is indicative of the Early Devonian *Ozarkodina eurekaensis* conodont Biozone. However, it is not clear which horizon in the *Scyphocrinites* limestone succession of the Filon Douze section the sample of Belka *et al.* represents. On the basis of this evidence, I consider the boundary questionably to be at the base of the crinoidal calcarenite (bed C) that succeeds the uppermost *Scyphocrinites* horizon (bed S-II) at Filon Douze.

Thickness. Approximately 7 m.

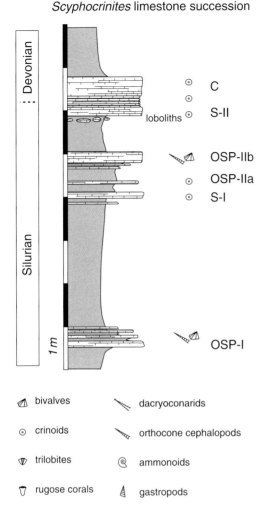

TEXT-FIG. 6. Section of the *Scyphocrinites* limestone succession, Přídolían–Lochkovian, Filon Douze, Tafilalt. Orthoceridans (mainly *Kopaninoceras* and *Plagiostomoceras*) are concentrated in beds OSP-I and OSP-II.

Description. The lowermost part of the succession (OSP-I) is a 0.5-m-thick succession of black biomicritic floatstone that is intercalated with thin argillaceous horizons, which contain masses of orthocones (*Kopaninoceras, Plagiostomoceras*) with cross-sections between 10 and 20 mm but without *Scyphocrinites*. They are strongly orientated in a north-east/south-west direction (50/230 degrees) with apices directed towards the south-west. At the base of the succession two individual 0.15-m-thick limestone beds are distinguishable.

Silt/claystones occur until *c.* 3 m above OSP-I where a 0.15-m-thick encrinitic packstone is intercalated (S-I). Pelmatozoan debris is the main bioclastic component of this dark limestone; orthocones are very rare.

Approximately 0.2 m above S-I a 0.05-m-thick grey biomicritic and partly biosparitic pack- to floatstone occurs (OSP-IIa) that contains masses of randomly orientated

small orthocones (cross-section *c.* 10–20 mm) and brachiopod shell hash. Several of the orthocone fragments are vertically orientated. Pelmatozoan ossicles are rare in OSP-IIa.

A black biomicritic floatstone (OSP-IIb), which contains common pelmatozoan debris and masses of orthocones, is intercalated within clay- to siltstones *c.* 0.5 m above OSP-IIa. The diversity of orthocones is exceptionally high. *Kopaninoceras, Orthocycloceras* and *Hemicosmorthoceras* are common, and a characteristic element is *Parakionoceras originale*; they are randomly orientated. The bed consists of a 0.25-m-thick conspicuous limestone at the top and a 0.25-m thickening-upward succession of argillaceous limestones at the base.

A 1-m-thick claystone occurs above OSP-IIb. Approximately 0.7 m from the base of this claystone there is a horizon that contains masses of articulated floats of *Scyphocrinites* ('loboliths'). The claystone is terminated by an encrinitic packstone (S-II) *c.* 0.5 m thick that contains commonly articulated parts of *Scyphocrinites*.

Above S-II is a massive crinoidal calcarenite 0.45 m thick (bed C). The top of bed C is a conspicuous, partly limonitically impregnated, omission horizon. No cephalopods occur in the lobolith horizon and in beds S-II and C.

The *Scyphocrinites* limestone succession is overlain by a ginger to pale grey succession of clays and siltstones more than 100 m thick. Approximately 40 m above, a horizon with common phosphatic nodules occurs (bed PK). These nodules contain masses of *Plagiostomoceras culter* and occasionally *Hemicosmorthoceras semimbricatum*.

3. Jovellania *limestone succession* (Text-fig. 7)

Biostratigraphy. Pedavis pesavis conodont Biozone, latest Lochkovian. The *Jovellania* limestone succession is not to be confused with the '*Jovellania*' Limestone', an informal term of H. Hollard that was adopted by Bultynck and Walliser (2000). The latter refers to a limestone that can be correlated with the *Deiroceras* Limestone of my investigation. In the *Deiroceras* Limestone of Filon Douze not a single *Jovellania* was found. In contrast, *Jovellania* and other oncocerids constitute a common element in the limestone succession described below. The *Jovellania* limestone succession is considered to represent the *pesavis* Bioevent (Lochkov/Prag Event *sensu* Schönlaub 1996, and Chlupáč and Kukal 1996), which is linked with a major world-wide regression (Talent *et al.* 1993; Hladíková *et al.* 1997). Belka *et al.* (1999) and Bultynck and Walliser (2000) correlated the base of the Pragian in the eastern Tafilalt with a conspicuous colour change in the siltstones below the Pragian limestones, which occurs at the Filon Douze section just above the *Jovellania* limestone horizon. The *Jovellania* limestone succession corresponds to bed TM 212 of Hollard (1977), and the occurrence of the tentaculite *Homoctenowakia bohemica* in it, and in bed BS, are indicative of the late Lochkovian age of the succession.

Thickness. Approximately 7 m.

Description. The lowermost bed, OJ-I, is a dark biomicritic silty floatstone <0.1 m thick that contains masses of poorly orientated orthocones of mainly *Adiagoceras* gen. nov. and *Hemicosmorthoceras*. Large bivalves (*Panenka*) occur at the top of the bed. The bed is characterized by the occurrence of scour and fill structures *c.* 0.15 m thick with a diameter of <0.5 m. These structures contain a floatstone with masses of poorly orientated orthocones. Beneath the small orthocerids, large specimens of *Sichoanoceras* and *Temperoceras*, and oncocerids such as *Jovellania* and *Bohemojovellania* are common.

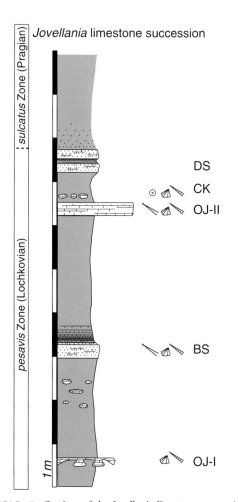

TEXT-FIG. 7. Section of the *Jovellania* limestone succession (base of OJ-I to top of DS), Lochkovian, Filon Douze, Tafilalt. This section is rich in orthoceridans (mainly *Adiagoceras* gen. nov. and *Hemicosmorthoceras*) and bivalves; jovellanids are common; for explanation of symbols, see Text-figure 6.

Above OJ-I a dark grey claystone follows that contains carbonate nodules up to 0.5 m wide which commonly contain orthocones. The claystones containing the nodules are topped by black bituminous argillaceous silt- to sandstone of 0.3 m (BS). Towards the top of BS the arenitic content decreases markedly and the bed appears as finely laminated, dark brownish, bituminous sandy shale. It commonly contains small fragments (cross-section <10 mm) of orthocones (*Arionoceras* and diverse sphaerorthocerids) and dacryonocarids (*Homoctenowakia bohemica*), which are the only macrofaunal components. The shales successively change towards dark grey clay–siltstones.

Approximately 3 m above BS a massive dark biomicritic floatstone 0.3 m thick occurs, which contains masses of very poorly sorted and orientated orthocones (OJ-II) in which dacryoconarids (*Homoctenowakia bohemica* and *Paranowakia intermedia*) are common. This bed is sorted with large components at the base and smaller components concentrated at the top. The largest fragments of orthocones are of *Temperoceras* with a cross-section diameter of more than 150 mm; the most frequent is *Adiagoceras* gen. nov. In the top 0.05 m of the bed masses of small bivalves (*Pterinea*) occur. This semi-infaunal bivalve is indicative of poorly oxygenated bottom conditions similar to those indicated by the closely related pectinids (Kříž 1999).

Directly above OJ-II there is a conspicuous purple to pinkish claystone that contains masses of pelmatozoan debris and small bivalves (*Cheiopteria*). Approximately 0.2 m above OJ-II a horizon is intercalated in this claystone with ochre-coloured limestone nodules that contain masses of small (cross-section 3 mm) *Cheiopteria* and commonly *Hemicosmorthoceras* (bed CK).

The top of the *Jovellania* limestone succession is a sandstone 0.5 m thick with a 0.1-m-thick arenitic marl intercalation in its middle part (DS). This sandstone is overlain by a yellowish, greenish to pale grey clay siltstone, which contains a sand fraction that successively decreases above its base.

4. *Pragian–Zlíchovian limestone succession* (Text-fig. 8)

Biostratigraphy. Eognathodus sulcatus conodont Biozone, earliest Pragian–*Polygnathus laticostatus-inversus* conodont Biozone, latest Zlíchovian, early Emsian (after Klug 2002). The ammonoid fauna of the upper part of the succession was described in detail by Klug (2002). The limonitic faunas of the claystones below the *Erbenoceras* and *Deiroceras* limestones are the subject of a monograph by Klug *et al.* (2008). However, despite the good overall fossil content, pinpointing the Pragian/Zlíchovian and the Zlíchovian/Dalejan boundaries at Filon Douze is not possible because index fossils that are used in Bohemia (see Chlupáč

1999) are rare around the potential boundary intervals. The earliest trilobites indicative of Pragian age are found at the top of bed KMO-I (*Crotalocephalina gibba*, *Reedops* sp.).

Thickness. Approximately 60 m.

Description. Directly above the uppermost sandstone bed (DS) of the *Jovellania* limestone succession a yellowish siltstone 2.5 m thick forms the basal Pragian. Within this siltstone (*c.* 0.05 m thick) layers of nodular limestones are intercalated that increase in thickness and carbonate content towards the top. The top of the layer is composed of a 0.2-m-thick, massive yellowish biodetrital succession of limestones with a high content of pelmatozoan debris, and fragments of trilobites (*Crotalocephalina gibba*, *Reedops* sp.; bed KMO-I). The dominant orthocones are *Temperoceras*, *Arthrophyllum* and *Spyroceras*.

Above these limestones, a 13-m-thick succession of platy greenish claystones follows; this is terminated by a massive greenish–pale grey biodetrital limestone 0.1 m thick, which contains pelmatozoan debris but otherwise is poor in macrofossils except for solitary corals (K4).

Limestone K4 is covered by a 5-m-thick succession of platy greenish claystones (KMO-II) intercalated with nodular pelmatozoan biodetrital limestone layers. These limestone layers are concentrated in the middle interval of KMO-II and contain a rich fauna of orthocones (predominantly *Arthrophyllum*, *Temperoceras* and pseudorthocerids). Trilobites, gastropods and solitary corals constitute a rich benthos.

Limestone K3 consists of two distinct pelmatozoan biodetrital wackestone to grainstone beds; the lower bed is 0.1 m and the upper bed 0.3 m thick; in between is a marl 0.2 m thick that contains limonitized pyrite crystals but lacks a macrofauna. The lower limestone bed of K3 is nodular and contains a high percentage of sand and biodetritus; the bivalve *Panenka* occurs occasionally. The upper limestone bed is massive; occasionally, large orthocerids occur (*Temperoceras*). The top of K3 is a conspicuous 0.05-m-thick horizon of a limonite-encrusted nodular limestone that is interpreted as an omission horizon.

Above K3 a claystone succession *c.* 8 m thick occurs (KMO-III), which in its basal part contains a considerable percentage of sand; towards the middle is an interval with numerous thin nodular limestone horizons and small favositid biostromes <0.2 m thick and *c.* 2–3 m in diameter. This interval is highly fossiliferous, containing mainly trilobites (*Gravicalymene* sp., *Reedops intermedius*, *Reedops* ex. gr. *cephalotes*, *Paralejurus* sp.). The cephalopod assemblage is dominated by pseudorthocerids, *Temperoceras* and *Arthrophyllum*.

Limestone K2 is a 0.2-m-thick, massive, biodetrital dacryoconarid grainstone, containing more biodetritus

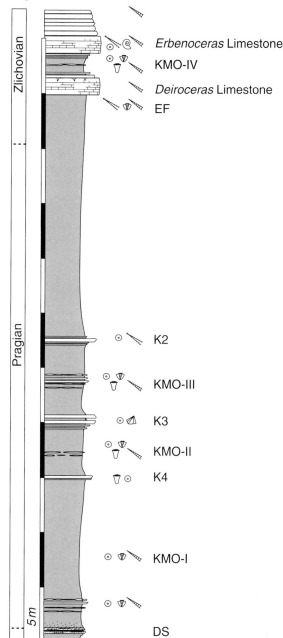

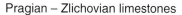

TEXT-FIG. 8. Section of the Pragian and Zlíchovian limestones of Filon Douze, Tafilalt. The first ammonoids occur in the *Erbenoceras* Limestone. The earliest occurrences of *Bactrites* and *Centroceras* are in beds KMO-III and KMO-IV respectively; for explanation of symbols, see Text-figure 6.

(mostly pelmatozoan debris) at its base. The dacryonoconarids (*Nowakia acuaria*) are deposited in cone-in-cone position, but macrofossils are generally very rare.

Above K2 is a monotonous greenish claystone 23 m thick in which no macrofossils were found apart from in the uppermost part, in a horizon *c.* 0.5–2 m below the

Deiroceras Limestone, where there is a conspicuous concentration of limonitized macrofossils (bed EF). This horizon is identical to Emsian faunule A horizon of Klug *et al.* (2008). The fauna consists predominantly of *Devonobactrites obliquiseptatum*, trilobites and hyolithids, and occasionally orthocerids.

The *Deiroceras* Limestone is a massive greyish wackestone 1.5 m thick that contains large (cross-section *c.* 50–100 mm) fragments of *Deiroceras*, and rare *Devonobactrites*. The orthocones are horizontally bedded, but have no preferred orientation. Often the conchs are partly dissolved and destroyed at the top, and locally winnowing of the micritic matrix has occurred, indicating reduced sedimentation rates. At the top of the *Deiroceras* limestone there is a limestone bed 0.02 m thick that displays an irregular limonitized, partly phosphoritized surface that is interpreted as an incipient hardground.

Above the *Deiroceras* Limestone there is a 2.5-m-thick succession of claystones with intercalated nodular bioclastic limestone beds (KMO-IV) that contain a rich fauna of rugose corals and trilobites. The cephalopods are less common than in the marls below; *Temperoceras* and *Arthrophyllum* are most frequent. At the top of this interval the carbonate content increases successively and merges with that of the overlying *Erbenoceras* Limestone.

The basal 0.3 m of the *Erbenoceras* Limestone consists of argillaceous nodular limestones that contain abundant large specimens of *Panenka*. Towards the top, there is a more massive limestone bed 1.5 m thick containing abundant orthocerids. The agoniatitid *Erbenoceras advolvens* is a conspicuous element of this horizon. The entire *Erbenoceras* Limestone is about 4.5 m thick, with an increasing number of fine argillaceous layers towards the top. The limestones are nodular, biodetrital wacke-packstones containing predominantly pelmatozoan debris and dacryoconarids. Klug (2002) reported *Nowakia zlichovenis magrebiana*, *N. praesulcata*, *N.* cf. *kabylica* and the trilobite *Sculptoproteus magrebus* from it. Cephalopods are common in the upper part of these limestones; orthocones such as *Metarmenoceras*, *Subormoceras* gen. nov. and *Devonobactrites* are most frequent. Large fragments of *Metarmenoceras* with cross-sections of 150–200 mm occur in the uppermost interval of the limestones together with abundant bivalves (predominantly *Panenka*).

The *c.* 125-m-thick Emsian claystones that overlie the *Erbenoceras* Limestone are poorly exposed at Filon Douze.

5. *Early Eifelian limestone succession*

Biostratigraphy. Late *Pedavis patulus* conodont Biozone–early *Oneotodus costatus* conodont Biozone, early–middle Eifelian (after Klug 2002). The ammonoids *Foordites ven-iens* and *Mimotornoceras djemeli* characterize the lower part of the interval representing the Choteč Event level (after Chlupáč 1985). The ammonoid *Pinacites jugleri* occurs just above the Choteč Event, providing evidence of the *jugleri* Event of Walliser (1985). The entire interval in the Tafilalt was part of a detailed stratigraphical investigation by Klug (2002).

Thickness. Approximately 2 m.

Description. The dark claystones and nodular limestones of the Choteč Event level (Bed F) mark the onset of a basic change in the sedimentological regime at Filon Douze and in the eastern Anti-Atlas Region inclusive of the Tafilalt. Above this level the predominantly siliciclastic succession with a few intercalated limestone horizons ('Amerboh Group' of Hollard 1981) changes towards mainly limestones with thin intercalated marls ('Bou Tchrafine Group' of Hollard 1981).

Bed F is composed of a dark, nodular, dacryoconarid packstone that is intercalated with irregular argillaceous layers. The entire horizon is 0.6 m thick and successively changes towards a more massive limestone at the top; its dominant macrofossil is the bivalve *Panenka*, which reaches 50 mm in size. The packstones and argillaceous interbeds contain masses of dacryoconarids that are often deposited in cone-in-cone position. Most nautiloids have imploded phragmocones with broken septa retained in the conchs. Ammonoids are common but usually fragmentary. They are dominated by agoniatids. Klug (2002) reported five agoniatitid species in bed F from the Jebel Ouaoufilal East. In addition to the ammonoids, *Neocycloceras* and oncocerids are conspicuous elements of bed F.

Above this bed is a 1-m-thick, massive, dark grey dacryoconarid packstone–floatstone that consists of two conspicuous, exceptionally fossil-rich layers at 0.2 m (PI) and 0.7 m (PII) from the base. Klug (2002) reported that the horizon is composed of two conspicuous fossiliferous layers throughout the Tafilalt and noted that several iron-rich bands occur in the limestone, and a sometimes winnowed micritic matrix between the biodetrital components indicates condensed sedimentation; ammonoids are the dominant macrofossils. Klug (2002) reported eight species of agoniatitids and four species of anarcestids from the Jebel Ouaoufilal East section, and emphasized that the ammonoid layers are composed exclusively of either anarcestids or agoniatitids. Bactritoids are the next most common cephalopods in PI and PII; slender orthocerids such as *Infundibuloceras* occur. The pseudorthoceratid *Subormoceras* is relatively common. The bactritoids and orthocerids are clearly orientated in a north-east–south-west direction. A dominant direction of the apical position was impossible to detect at Filon Douze, rendering the origin of the currents somewhat uncertain. A rich benthic fauna of gastropods, trilobites and bivalves characterizes this limestone bed.

INTERPRETATION OF THE SEDIMENTARY SUCCESSION

General remarks

The sedimentary succession is dominated by thick claystones, which are intercalated by horizons of limestones, siltstones and sandstones. It represents a mixed carbonate-siliciclastic system on the continental shelf of northern Gondwana. The lack of turbidites is interpreted as evidence for a very gently sloping sea-floor. A predominantly E/W–NE/SW orientation of orthocones reflects a shoreline-parallel current regime at times of orthocone deposition (compare with the palaeogeographical reconstructions of Legrand 2003).

The limestone and sandstone horizons, which are intercalated within the claystones, mark the boundaries of five main depositional cycles. At the base of each cycle arenitic and tempestite horizons occur; the top of some cycles show massive limestone beds. Quartz arenites, siltstones and tempestites are interpreted as representing lowstands. Sideritic and phosphoritic nodular horizons within the claystones (bed PK) and claystones containing secondarily limonitized pyrite nodules and limonitized fossils are interpreted as representing a reduced sedimentation rate when sea-level rise was rapid during transgressions. Generally, the claystone beds contain a higher silt fraction at the base of each sedimentary cycle.

No continuous trend of deepening or shallowing of the depth of deposition through the succession can be detected. Sandstones occur at the top of the *Temperoceras* and *Jovellania* limestone successions, and some horizons within the Pragian limestone succession contain a considerable percentage of quartz sand. Proximal tempestites occur within the *Temperoceras*, *Scyphocrinites*, *Jovellania*, and Pragian and Zlíchovian limestone successions. Deep erosional structures at the base of the pre-Pragian tempestites and their coarser components indicate a higher energy regime during deposition, which is attributed to their generally shallower depths of deposition compared to the Pragian–Eifelian beds. By contrast, the post-Lochkovian sediments (including the tempestites) contain a considerably richer benthos, indicating shallower depositional conditions than the Ludlow–Lochkovian sediments, and the occurrence of small favositid bioherms within Zlíchovian marls and claystones indicates deposition within the photic zone. However, the lack of benthos, except for epibyssate bivalves within pre-Pragian sediments, is not a sufficient argument for deeper depositional settings; I interpret it as evidence for reduced bottom oxygenation.

In conclusion, the sedimentary succession of the Filon Douze section is consistent with the eustatic curve presented by Johnson *et al.* (1985) and Buggisch and Mann (2004, fig. 13) with major sequence boundaries around the Silurian/Devonian boundary, the Lochkovian/Pragian boundary, and in the upper Dalejian.

Description of depositional cycles

The lowermost cycle of the measured section is the massive limestone of bed OP and the overlying tempestite bed OP-K, which is interpreted as representing a shallowing sequence of the highstand systems tract (Text-fig. 9).

The base of the subsequent cycle is placed at the base of the sandstone at the top of the *Temperoceras* limestone succession. The subsequent transgression of the second cycle is recorded by a decrease in sand and silt content of the claystones. An interval within the claystone, *c.* 30 m above the base of the sandstone, commonly containing sideritic nodules, is interpreted as a maximum flooding horizon. The *Scyphocrinites* limestone succession, which is composed of proximal tempestites and calcarenites, is interpreted as a shallowing-up sequence representing the highstand systems tract of the second cycle. The top of this shallowing-upwards sequence is the crinoidal calcarenite bed C, which is regarded as the base of the third cycle and correlates with cycle Ia$_2$ of Buggisch and Mann (2004).

The transgression of the third cycle is recorded by a thick succession of partly well-laminated claystones. The nodular bed PK within these claystones is interpreted as a maximum flooding horizon. The tempestites and argillaceous sandstones of the *Jovellania* limestone succession are interpreted as representing the progradation of clastic sediment within the highstand systems tract of the third cycle.

The fourth cycle, which can be correlated with depositional cycle Ia$_1$ of Buggisch and Mann (2004), begins at the base of sandstone DS at the top of the *Jovellania* limestone succession. The decreasing quartz and silt content of the subsequent marl-limestone alternations indicates a deepening of the depositional setting. Limestone K3, which shows a conspicuous omission horizon at the top and contains a considerable fraction of sand, is interpreted as the base of the fifth cycle.

Within the fifth cycle, which can be correlated with cycles Ib and Ic of Johnson *et al.* (1985) and Buggisch and Mann (2004), four subordinate cycles can be distinguished. The lowermost subcycle consists of a thick claystone that contains at its top a conspicuous limonitized fauna (bed EF) and limonite nodules. Bed EF is interpreted as the maximum flooding horizon of this subcycle. The overlying *Deiroceras* Limestone shows evidence of

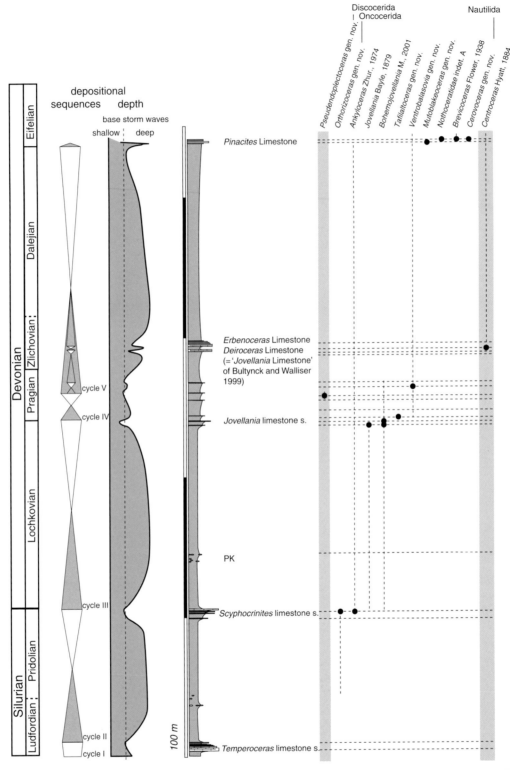

TEXT-FIG. 9. The complete section at Filon Douze showing limits of depositional sequences, reconstructed depositional depth, and stratigraphical ranges of the non-ammonoid cephalopods illustrated.

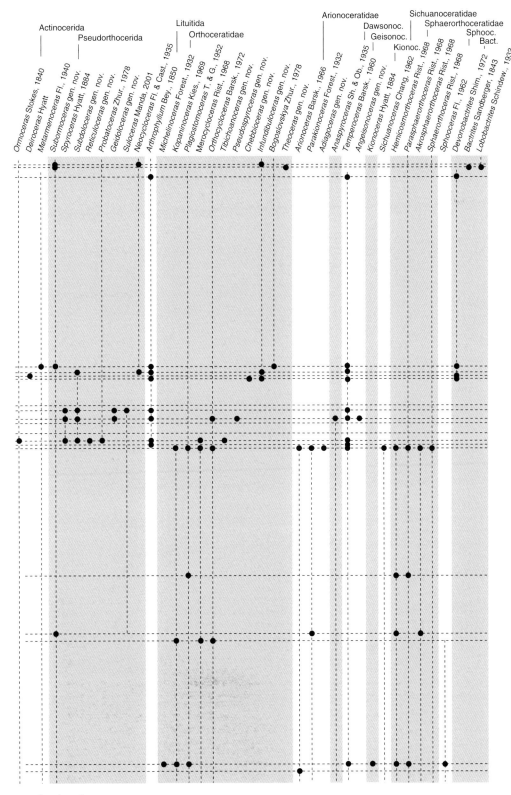

TEXT-FIG. 9. Continued.

reduced sedimentation rates and a conspicuous hard-ground at the top. This hardground is interpreted as the boundary with the second subcycle. The second subcycle consists of several metres of claystones at its base, including small *Favosites* bioherms, which merge smoothly towards nodular limestones at the base of the *Erbenoceras*

Limestone. The overlying middle part of this limestone is a condensed biodetrital wackestone with several inconspicuous omission horizons. The boundary with the uppermost subcycle of the fifth cycle is placed at the middle of the *Erbenoceras* Limestone, which towards its top merges into a nodular argillaceous wackestone. A thick succession of claystones marks the lower part of the uppermost subcycle of the fifth depositional cycle. The carbonate content of these claystones increases towards the top and the higher part of the beds is composed of an alternation of nodular limestones and claystones. The dark nodular dacryoconarid packstone of the transgressive Choteč event (Klug 2002) terminates the fifth cycle.

CEPHALOPOD SUCCESSION

Diversity signal

On average the cephalopod-bearing beds of the Filon Douze section each contain seven species belonging to six genera (ammonoids excluded). The *Deiroceras* limestone is the only monospecific bed, containing exclusively *Deiroceras hollardi* gen. et sp. nov. The richest cephalopod fauna recovered, containing a total of 17 species belonging to 13 genera, is from the tempestite bed OJ-I. The results of standardized diversity estimation show exceptionally low values for the Chao I richness estimator for beds PK ($S_{chao1} = 4$) and EF ($S_{chao1} = 5$) (Text-fig. 10; Table 1). The associations of both beds are preserved in clay-siltstones and are interpreted as representing the deepest depositional environments of the Filon Douze section (see above). In bed PK the cephalopod fauna is dominated by *Plagiostomoceras*, in bed EF by *Devonobactrites*. Both low diversity values are supported by a high number of samples. In contrast, the highest estimated Chao 1 richness occurs in the Pragian marl KMO-III ($S_{chao1} = 25$). However, the richness is only slightly lower in other Pragian marls and in the tempestitic coquininas OP-M ($S_{chao1} = 21$) and OJ-I ($S_{chao1} = 18$). There is no general trend of increasing or decreasing richness within the observed interval.

The measure of taxonomic distinctness is independent of sample size. Therefore, those beds containing more than two species of cephalopods are sufficient for taxonomic distinctness calculations. The results show that a significant increase in taxonomic distinctness values from $\Delta^+=2.5$ towards values of $\Delta^+>3.25$ occurs with the onset of the *Jovellania* limestone succession (Text-fig. 10; Table 4). Whilst the *Temperoceras* and *Scyphocrinites* limestone successions contain almost exclusively Orthocerida and rarely Pseudorthocerida or Oncocerida, the *Jovellania* limestone succession and higher beds are characterized by an additional modest content of Discosorida, Lituitida,

Bactritida and Nautilida. The taxonomic distinctness is not a proxy of disparity, because genera such as *Arthrophyllum* and *Bactrites* are orthocones, showing similar morphology to other orthocones of the Orthocerida, but represent different higher taxa. Therefore, the taxonomic distinctness provides no estimate of the disparity of the cephalopod fauna. However, the occurrence of brevicones and coiled nautiloids in the post-Lochkovian strata of the Filon Douze section is evidence for an increased disparity in the Pragian–Middle Devonian compared with the Silurian and lowest Devonian. Remarkably, this increase in disparity and taxonomic distinctness occurs below the significant Lochkovian–Pragian facies change at Filon Douze. Because no data from other localities of that time interval are available, it is impossible to compare the diversity trends with a supra-regional pattern.

Besides the low evenness values of beds PK and EF no trend in change of evenness can be detected in the section. The low evenness values of beds PK and EF reflect the strong dominance of *Plagiostomoceras* and *Devonobactrites* respectively; it is interpreted as reflecting the deep depositional environment of these beds. Accordingly,

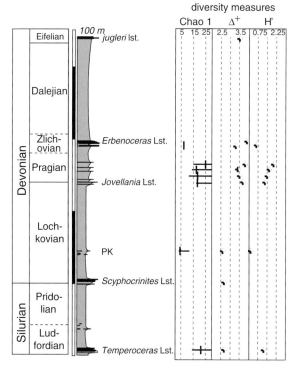

TEXT-FIG. 10. Diversity measures of selected beds in the Filon Douze section. $S_{Chao\ 1}$, species richness after Chao (1984) with bars indicating 95 per cent confidence level; Δ^+, taxonomic distinctness after Clarke and Warwick (1998, 1999); H', Shannon index. Note the distinct increase in all values within and above the *Jovellania* limestone succession. Very low diversity and evenness values occur in bed PK and EF. For data, see Table 1.

these two genera are interpreted as being orthocones that inhabited distal environments, probably in a pseudo-planktonic lifestyle.

Morphological signal

Orthocones dominate the cephalopods in the succession. Two groups can be distinguished. The first comprises those with a comparatively high angle of expansion, wide siphuncle that may be expanded within the chambers, and massive cameral and endosiphuncular deposits; representatives are the Actinoceratida, Pseudorthoceratida, and Geisonoceratidae such as *Temperoceras*. *Arthrophyllum* also belongs here (Table 4). I include these forms within the euorthocones (new term). The second group comprises slender, often small conchs, with a narrow siphuncle, often having wide septal spacing. Cameral and endosiphuncular deposits are suppressed or lacking in these conchs. Representatives are Arionoceratidae, Orthoceratidae and bactritoids, which I include in the angustocones (new term). *Plagiostomoceras* is assigned to the angustocones, although the type species has cameral deposits. This is justified as all the species occurring in the Filon Douze section have slender shells with a narrow siphuncle and without deposits. Their narrow siphuncle indicates a low cost for buoyancy regulation and suggests an energy-intensive lifestyle (see Kröger 2003). The lack of cameral deposits is probably an effect of a reduced liquid passage through the phragmocone during life. They are interpreted as cephalopods with low energy needs adapted to environments with low food availability, which lived as vertical migrants in the free water column. As the distal environments of the free water column are relatively poor in food availability, angustocones are interpreted as planktonic dwellers.

In contrast, euorthocones are interpreted as cephalopods with a comparatively energy-intensive lifestyle; they were potentially also active vertical migrants. Increased internal cameral surfaces in many forms and the comparatively large siphuncular surface are evidence for an ability actively to regulate buoyancy. However, buoyancy regulation of the euorthocones was much more cost-intensive than that in angustocones, demanding more active searching for high energy food. The increased siphuncular surface must have caused comparatively high-cost buoyancy regulation (see Kröger 2003). A demersal lifestyle of euorthocones, probably comparable to that of the recent *Nautilus,* is inferred from this evidence.

When comparing the relationship of the richness of these two groups in the individual beds at Filon Douze a clear pattern is visible (Text-fig. 11). The pre-Pragian

TABLE 4. Orthocone cephalopods of the Filon Douze section classified as angustocones or euorthocones. Euorthocones have comparatively large siphuncles, endosiphuncular and often heavy cameral deposits. Angustocones have narrow siphuncles, suppressed endosiphuncular and cameral deposits and often slender conchs with wide septal spacing.

Order	Genus	Angusto-cone	Euortho-cone
Actinocerida	*Metarmenoceras*		x
	Ormoceras		x
	Deiroceras		x
Pseudorthocerida	*Diagoceras*		x
	Geidoloceras		x
	Neocycloceras		x
	Probatoceras		x
	Reticuloceras gen. nov.		x
	Subdoloceras gen. nov.		x
	Subormoceras gen. nov.		x
	Spyroceras		x
	Cancellpyroceras gen. nov.		x
	Suloceras		x
Lituitida	*Arthrophyllum*		x
	Sphooceras		x
Orthocerida	*Chebbioceras*	x	
	Infundibuloceras	x	
	Kopaninoceras	x	
	Merocycloceras	x	
	Michelinoceras	x	
	Orthocycloceras	x	
	Plagiostomoceras	x	
	Pseudospyroceras gen. nov.	x	
	Theoceras gen. nov.	x	
	Tibichoanoceras gen. nov.	x	
	Arionoceras	x	
	Adiagoceras gen. nov.	x	
	Parakionoceras	x	
	Anaspyroceras		x
	Angeisonoceras gen. nov.		x
	Temperoceras		x
	Kionoceras	x	
	Sichuanoceras		x
	Sphaerorthoceras	x	
	Akrosphaerorthoceras	x	
	Parasphaerorthoceras	x	
	Hemicosmorthoceras	x	
Bacritida	*Bactrites*	x	
	Devonobactrites	x	
	Lobobactrites	x	

beds EF and F have very low percentages of euorthocones, whereas they are the dominant group in the Pragian–Middle Devonian beds. Bed EF is a clay-siltstone horizon containing a rich benthic and nectonic limonitized fauna including trilobites, gastropods and acanthodians (Klug *et al.* 2008). Bed F with a dark marl to nodular limestone lithology has masses of dacryoconarids, representing the Choteč transgressive event (see Klug 2002). Bed PK, which yields exclusively angustocones (*Plagiostomoceras culter*, *Hemicosmorthoceras semimbricatum*) is a limestone nodule horizon within platy claystones–siltstones that commonly contains graptolites underneath angustocone orthocones and a low diversity fauna of epibyssate bivalves. Beds PK, EF and F are interpreted as representing the deepest, most distal depositional settings of the section during maximum flooding intervals. The strong dominance of angustocones in these beds supports the suggested planktonic, energy-intensive lifestyle of these forms. A similar concentration of angustocones has been reported from distal environments in the Ludlow type area (Hewitt and Watkins 1980), where *Arionoceras gregarium* and *Hemicosmorthoceras dimidiatum* are the dominant taxa. Wilmsen (2006) reported the occurrence of frequency peaks of organisms adapted to low food availability during maximum flooding intervals within the boreal Upper Cretaceous limestones of central Europe. A similar faunal pattern occurs in the Siluro-Devonian strata of the Tafilalt.

Euorthocones are generally strongly under-represented in all pre-Pragian beds and the benthos indicates oxygen-poor conditions at the bottom which, however, are not entirely linked with depth of deposition. Although beds PK, EF and F represent comparatively deep conditions below storm wave-base, the fauna of the pre-Pragian tempestites and sandstones was deposited under shallow conditions above storm wave-base. The depositional depths recorded at Filon Douze are in good accordance with supra-regional sea-level curves (Johnson *et al.* 1985; Buggisch and Mann 2004) in which the Pragian eustatic sea level is not significantly lower compared with the Lochkovian. Nevertheless, tempestites and sandstones contain few euorthocones and a restricted benthos largely composed of epibyssate bivalves.

The faunal composition dramatically changes at the base of the Pragian above bed OJ-II. The entire upper part of the section is composed of a diverse benthos and a predominance of euorthocones, even in parts that are interpreted as deposited below storm wave-base, such as the marl beds KMO-I, II and III. This faunal change is accompanied by a marked change in sediment colour from black or dark brown to light brown or yellowish.

Dramatic faunal and sedimentary colour changes are not restricted to the Filon Douze section. Alberti (1980), Belka *et al.* (1999), and Bultynck and Walliser (2000, p. 216) recorded the disappearance of black sediments above the uppermost 'Orthoceras' Limestone in proximity to the Lochkovian/Pragian boundary in the northern Tafilalt. Kříž (1998, 1999) reported the disap-

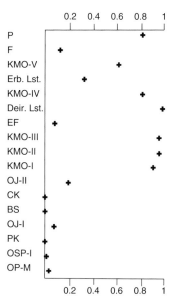

TEXT-FIG. 11. Ratio of species richness of euorthocone/angustocone orthocerids in selected beds of the Filon Douze section. Note the distinct increase in the ratio above the *Jovellania* limestone succession (bed OJ-II) and the low values for beds EF and F. For data, see Table 1.

pearance of the orthocone-bivalve community that is typical of Silurian–earliest Devonian successions worldwide at the end of the Lochkovian, and an increased benthic diversity and a change towards lighter sediment colours above the Lochkovian/Pragian has been reported from the Barrandian area (Chlupáč *et al.* 2000), the Carnic Alps (Schönlaub 1996) and the Iberian massifs (Garcia-Alcalde *et al.* 2000; Robardet and Gutiérrez-Marco 2004). A major carbon isotope shift was recorded by Buggisch and Mann (2004) at the Lochkovian/Pragian transition in central and southern Europe. Additionally, a clear shift from low diversity orthocerid-bivalve-dominated palaeocommunities towards palaeocommunities containing a rich benthos has been reported from Sardinia (Ferreti *et al.* 1999). A radiation of pseudorthocerids and non-orthocone, non-ammonoid cephalopods occurs in post-Lochkovian limestones in the Barrandian (Manda 2001). The trend visible at Filon Douze appears, therefore, to be representative of a wider range of peri-Gondwanan areas of the Palaeo-Tethys. The poorly oxygenated bottom conditions even in shallow-water settings that characterized long intervals in the Silurian and earliest Devonian disappeared suddenly in the Pragian and culminated in the establishment of carbonate platforms in the Emsian and Middle Devonian. At Filon Douze a change in the non-ammonoid cephalopod fauna corresponds to this trend, leading from faunas dominated by angustocones to those dominated by euorthocones.

Evolutionary signal

The top of the *Jovellania* limestone succession, which marks the Lochkovian/Pragian boundary, coincides with the disappearance of a number of characteristic Silurian cephalopods at Filon Douze. *Hemicosmorthoceras, Akrosphaerorthoceras, Arionoceras* and *Parakionoceras* have their last occurrence in the limestone, and are not reported to occur in younger beds elsewhere. The subsequent Pragian strata contain a cephalopod fauna that is characterized by a high number of endemic genera, and genera known exclusively from the late Early Devonian and early Middle Devonian, such as many pseudorthocerids and *Arthrophyllum*.

A strong evolutionary pulse occurs in the Zlíchovian, in which many new genera first occur and major new groups, such as bactritoids, ammonoids and nautilitids, suddenly appear. Bactritoids appear at Filon Douze in the lower Zlíchovian (*Devonobactrites obliqueseptatus*). Only a single questionable fragment of a bactritoid was found in pre-Emsian strata (KMO-III, Pragian). This local result reflects a general observation. Despite intensive collecting effort and repeated monographic treatment of Siluro-Devonian cephalopod faunules, no additional bactritoid specimens have been reported in pre-Emsian strata since the work of Ristedt (1981). A single poorly preserved and insufficiently illustrated bactritoid was reported by Termier and Termier (1950, pl. 87, figs 27–29) from the Upper Silurian of the Tafilalt east of Erfoud. Holland (2003) noted that this specimen should at best be disregarded. Thus, it appears that bactritoid orthocones are very rare or absent in pre-Emsian strata with the exception of '*Bactrites*' *bohemicus*. However, although assigned to *Bactrites*, the Ludlovian '*B.*' *bohemicus* was regarded as an early Zlíchovian homoeomorph of the Devonian *Bactrites* by Ristedt (1981). New data of the apical morphology of *Devonobactrites* from the lower Zlíchovian at Filon Douze support his opinion. The protoconch of *Devonobactrites* is elliptical, egg-shaped, and the conch adoral of the apical chamber is slender and displays a cross-section diameter less than that of the protoconch. This morphology is similar to the apex known from *Bactrites gracile* (Blumenbach, 1803) (= *Orthoceras schlotheimii* Quenstedt, 1845; compare Kröger *et al.* 2005). Therefore, the apical shape of *Bactrites gracile* differs from that of '*B.*' *bohemicus* in which the protoconch is subspherical and has a smaller diameter than the subsequent conch section. Moreover, the protoconch of '*B.*' *bohemicus* is about 0.2 mm in length and cross-section whereas that of *Devonobactrites obliquiseptatus* is 0.8 mm long and 0.4 mm in diameter. The dimensions of the latter are similar to those of protoconchs of Middle Devonian bactritoids (see Erben 1960) and early ammonoids (see Erben 1964). Thus, the available data show that Devonian bactritoids have exclusively an elliptical protoconch apical of a slender conch, which over a significant initial distance is narrower in cross-section than the relatively large protoconch (cross-section *c.* 0.5 mm). '*Bactrites*' *bohemicus* contrasts markedly with these later forms and is more similar in apical shape to the mid Silurian *Sphaerorthoceras beatum, S. carnicum* and *S. teicherti* (all Ristedt, 1968). Instead, the apex of the Devonian bactritoids strongly resembles the apices of the Middle Silurian '*Sphaerorthoceras*' spp. A–C and E–F described by Ristedt (1968) and the Early Devonian *Protobactrites*? sp. (Serpagli and Gnoli 1977, text-figs 5–6). These latter forms have a protoconch that is larger in cross-section than the subsequent slender conch. The commonality of apical shape, apical dimension and eccentricity of the siphuncle in the case of *Protobactrites* is a strong argument for a phylogenetic lineage leading to the Devonian bactritoids (Kröger and Mapes 2007). Therefore, an origin of bactritoids not earlier than Pragian is highly probable.

Beneath the first appearance of bactritoids in the Pragian and ammonoids in the Zlíchovian, the first tightly coiled representative of the Nautilitida is *Centroceras* sp. in bed KMO-IV in the Zlíchovian. The occurrence of *Centroceras* at Filon Douze is the earliest occurrence of a tightly coiled nautilid anywhere; it marks the initial appearance of a group of coiled nautiloids that dominated the Late Palaeozoic and persists with Recent *Nautilus*. The phylogeny of early nautilids is problematic; Dzik and Korn (1992) considered openly coiled Pragian oncocerids or trochospiral lechritrochocerids as possible ancestors. The first tightly coiled planispiral nautiloids are now known from the Early Emsian.

In conclusion, the stratigraphical pattern apparent at the Filon Douze section indicates an evolutionary pulse with the onset of several high level taxa in the late Pragian–Early Emsian. It occurs considerably above the major facies change at the base of the Pragian.

SYSTEMATIC PALAEONTOLOGY

Institutional abbreviations. MB.C., Museum für Naturkunde Berlin; PIMUZ, Paläontologisches Museum and Institut, Universität Zürich.

Class CEPHALOPODA Cuvier, 1797
Order DISCOSORIDA Flower, *in* Flower and Kummel 1950
Family PHRAGMOCERATIDAE Miller, 1877

Genus PSEUDENDOPLECTOCERAS gen. nov.

Type species. Pseudendoplectoceras lahcani sp. nov., from the Pragian of the Filon Douze section.

Derivation of name. Greek, *pseudes*, false, because the new genus resembles *Endoplectoceras*.

Other species included. None.

Diagnosis. Trochoceroid phragmoceratid with circular or slightly compressed cross-section and marginal siphuncle on concave side of shell curvature; angle of expansion *c.* 15 degrees; shell surface with faint straight, transverse growth lines and very shallow irregular undulations; siphuncle large with diameter 0.3 of conch cross-section; nummoidal siphuncular segments weakly expanded within chambers; adoral parts of siphuncular segments more strongly expanded than adapical parts; endosiphuncular deposits form adapically elongated bullettes.

Remarks. This genus is similar to the mid Silurian *Endoplectoceras* Foerste, 1926, in general conch shape, but differs in having more strongly expanded siphuncular segments with a simple convex connecting ring. The conch of *Endoplectoceras* differs further in being more compressed and less strongly coiled.

Occurrence. Pragian; Morocco.

Pseudendoplectoceras lahcani sp. nov.
Plate 1, figures 1–4; Plate 16, figures 9–11

Derivation of name. In honour of Filon Douze miner Lahcan Caraoui.

Holotype. MB.C.10184 (Pl. 16, figs 9–11).

Type locality and horizon. Filon Douze section, bed KMO-II, Pragian.

Other material. 9 fragments from the type locality and horizon.

Diagnosis. As for genus.

Description. Conch trochoceroid, lateral shift of coiling axis of <5 mm across coiling of 180 degrees in MB.C.10185 with smaller diameter 10 mm; larger diameter *c.* 24 mm and length of coiling axis *c.* 49 mm. Largest fragment, MB.C.10187, with dorsoventral cross-section of *c.* 40 mm. Conch circular or slightly compressed with ratio of conch width to height *c.* 0.9. Conch surface ornamented with distinct transverse, irregularly but narrowly spaced growth lines and shallow, directly transverse, irregularly spaced undulations. Septal distance at conch diameter of 25 mm, 2.5 mm. Septal perforation marginal with elliptically compressed cross-section; 3.8 mm wide and 5 mm high at conch diameter of 21.7 mm in MB.C.10188. Septal necks strongly recumbent, touching adapical surface of septa at side of septum, directed towards conch centre, loxochoanitic on siphuncular side, directed towards conch margin

(Pl. 16, figs 9–11). Siphuncular segments only slightly expanded. Siphuncle more expanded at adoral part of each segment. Endosiphuncular deposits elongated in apical direction.

Occurrence. As for genus.

Order ONCOCERIDA Flower, *in* Flower and Kummel 1950
Family BREVICOCERATIDAE Flower, 1945

Genus BREVICOCERAS Flower, 1938

Type species. *Brevicoceras casteri* Flower, 1938, from the Windom Member, Moscow Formation, Middle Devonian, of Filmore Glen, Moravia, New York, USA.

Diagnosis (from Flower 1938, p. 24). Slightly cyrtoconic brevicones with depressed cross-section; in early growth stages prosiphuncular side slightly more rounded than dorsum; near base of mature living chamber antisiphuncular side flattened; near aperture cross-section rounded-subtriangular with flattened antisiphuncular side; in dorsoventral direction prosiphuncular side convex, antisiphuncular side adapically concave, convex until base of mature body chamber, straight or only faintly convex until mature peristome; adjacent body chamber straight or faintly concave, close to peristome; position of largest gibbosity varies from species to species but generally at base of body chamber to adoral part of phragmocone in mature specimen; septa shallow and evenly curved; sutures sinuous with broad shallow lobes, lateral saddles and conspicuous lobe on antisiphuncular side; siphuncle nummoidal; actinosiphuncular deposits occur.

Occurrence. Middle–Late Devonian; Morocco, North America and Russia.

Brevicoceras magnum sp. nov.
Plate 1, figures 5–7; Text-figure 12A–C

Derivation of name. Latin, *magnus*, large, referring to the adult size of this *Brevicoceras*.

Holotype. MB.C.9632 (Pl. 1, figs 5–7).

Type locality and horizon. Filon Douze section, bed PI, early *O. costatus* conodont Biozone, middle Eifelian.

Other material. Known only from the holotype.

Diagnosis. *Brevicoceras* with adult size of *c.* 40 mm in conch height; shell surface faintly undulating, with conspicuous, irregularly spaced growth lines; cross-section

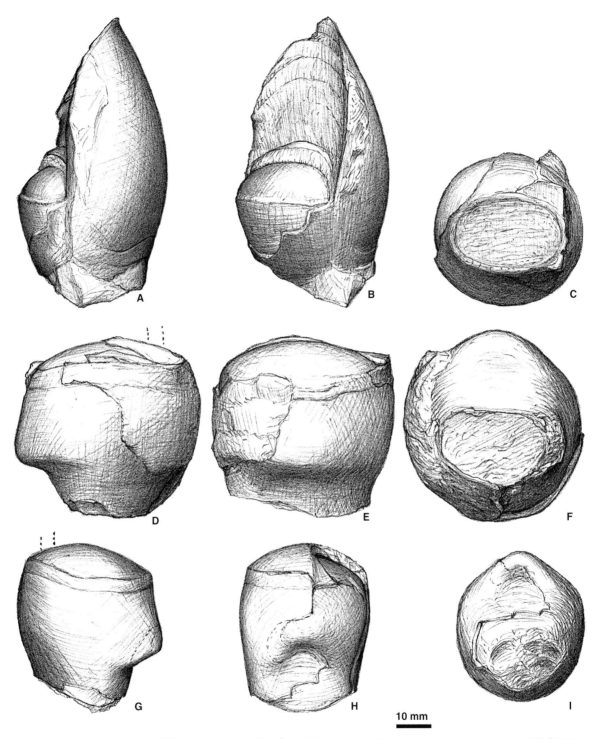

TEXT-FIG. 12. Reconstructions of Eifelian Brevicoceratidae from Filon Douze. A–C, *Brevicoceras magnum* sp. nov.; MB.C.9632. A, lateral, B, ventral and C, adoral views. *Cerovoceras brevidomus* gen. et sp. nov.; MB.C.9620. D, lateral, E, ventral and F, adoral views. G–I, *Cerovoceras fatimcaraouii* gen. et sp. nov.; MB.C.9623. G, lateral, H, ventral and I, adoral views. Note the characteristic shape of the body chamber horn, and the general shape of the body chamber. Dashed lines mark position of siphuncle. Scale bar represents 10 mm. Drawings by E. Sievert.

slightly depressed; length of mature body chamber *c.* 35 mm and height *c.* 40 mm; aperture broadly elliptical with more rounded prosiphuncular side; adjacent body chamber contracted with straight or slightly concave conch margin on antisiphuncular side and convex on prosiphuncular side in lateral view.

Description. Conch surface slightly irregularly undulating with distinctive growth lines that are regularly spaced in adapical parts. Distance between growth lines 1 mm. Mature aperture preserved in fragments. Cross-section of aperture broadly elliptical; width 37 mm, height 24 mm; prosiphuncular side more rounded. Adoral 12 mm of body chamber contracted with slightly concave outline in dorsoventral view. At base of mould of body chamber numerous broad, shallow, longitudinal striae represent muscle impressions (Pl. 1, fig. 5). Septa strongly concave with curvature of 11 mm. Septal distance 9 mm at adoral part of phragmocone, 6 mm at apical part of fragment. Sutures form shallow lateral lobes.

Remarks. *B. casteri*, which is similar in adult size, differs in that the adoral part of the body chamber is longer and more elongated. Other species of *Brevicoceras* are smaller than *B. magnum*.

Occurrence. Middle Eifelian; Morocco.

Genus CEROVOCERAS gen. nov.

Type species. *Cerovoceras fatimi* sp. nov., from the middle Eifelian, Filon Douze section.

Derivation of name. Latin, *cerus*, horn, referring to the conspicuous horned body chamber and resemblance to *Ovoceras* Flower, 1936.

Other species included. *Cerovocerus brevidomus* sp. nov.

Diagnosis. Brevicones with conspicuous horn on antisiphuncular side of body chamber; cross-section subcircular, slightly depressed or compressed; body chamber shorter than wide; mature aperture rounded-subtriangular with flattened antisiphuncular side; it opens body chamber only at prosiphuncular half; antisiphuncular side shorter and carrying conspicuous shallow, subtriangular projection, or horn; in median section prosiphuncular conch side convex, antisiphuncular nearly straight or faintly convex at base of body chamber towards horn; at position of horn, body chamber constricted, forming wide, slightly elongated, apertural opening; position of largest gibbosity at base close to most adoral part of phragmocone in mature specimens; septa oblique, sloping in apical direction on prosiphuncular side; sutures straight; septal curvature shallow; siphuncle close to conch margin; septal perforation ovate, compressed in type species.

Remarks. The shape of adapical growth stages and characters of septal necks of *Cerovoceras* are not known. Its assignment to the Brevicoceratidae is justified by its strong similarity to other Brevicoceratidea such as *Ovoceras* and *Brevicoceras* Flower, 1936, from which it differs mainly in having a conspicuous subtriangular body chamber projection.

Occurrence. Eifelian; Morocco.

Cerovoceras fatimi sp. nov.
Plate 1, figures 9–10, 14; Text-figure 12G–I

Derivation of name. In honour of Fatim, son of Filon Douze miner Lahcan Caraoui.

Holotype. MB.C.9623 (Pl. 1, figs 9–10, 14).

Type locality and horizon. Filon Douze section, bed PI, early *O. costatus* conodont Biozone, middle Eifelian.

Other material. Known only from the holotype.

EXPLANATION OF PLATE 1

Figs 1–4. *Pseudendoplectoceras lahcani* gen. et sp. nov.; bed KMO-II, Pragian. 1–3. MB.C.10185. 1, adapical view, not whitened, showing the slightly asymmetrical position of the siphuncle. 2, lateral view. 3, apical view. 4, MB.C. 10186, lateral view.

Figs 5–7. *Brevicoceras magnum* sp. nov., MB.C.9632, holotype; bed PI, early *O. costatus* conodont Biozone, middle Eifelian. 5, lateral view of cast of body chamber showing longitudinal impressions of muscle scars. 6, adoral view, showing shape of aperture. 7, lateral view of body chamber.

Figs 8, 11. *Jovellania cheirae* sp. nov, MB.C.9603, holotype; bed OJ-II, Lochkovian. 8, ventral view of cast of adapical part of body chamber and phragmocone. 11, adapical view.

Figs 9–10, 14. *Cerovoceras fatimi* gen. et sp. nov., MB.C.9623, holotype; bed PI, early *O. costatus* conodont Biozone, middle Eifelian. 9, adoral view of body chamber, showing adult aperture and shape of the dorsal horn. 10, lateral view of complete body chamber, position of siphuncle at left side. 14, detail of angular impression between dorsal horn and aperture, adoral view with dorsal side up; scale bar represents 10 mm.

Figs 12–13. *Cerovoceras brevidomus* gen. et sp. nov., MB.C.9620, holotype; bed PI, early *O. costatus* conodont Biozone, middle Eifelian. 12, adoral view of body chamber, showing adult aperture and shape of the dorsal horn. 13, lateral view of complete body chamber, position of siphuncle at left side.

All specimens from the Filon Douze section and × 1 except where indicated by scale bar.

PLATE 1

KRÖGER, nautiloids

Diagnosis. Cerovoceras with adult size of *c.* 40 mm in conch height and compressed cross-section; ratio of conch width to height 0.9; shell surface smooth; length of mature body chamber *c.* 30 mm; aperture rounded-triangular, height *c.* 24 mm, width 27 mm; antisiphuncular side of aperture flattened; conspicuous subtriangular projection of body chamber on antisiphuncular side 17 mm adoral of last septum; projection forms triangular arc directed towards antisiphuncular side, with central subtriangular depression; septa oblique, sloping adapically on prosiphuncular side; siphuncle close to conch margin, compressed ovate in cross-section, with diameter of septal perforation 0.2 of conch cross-section.

Description. Holotype is fragment of complete body chamber and two latest chambers of phragmocone. Conch height at base of body chamber 35 mm, conch width *c.* 33 mm. Body chamber length from latest septum to peristome *c.* 30 mm; maximum gibbosity of body chamber occurs 17 mm adoral of base. Here on antisiphuncular side, conch projection occurs and body chamber constricted just adoral of projection forming rounded-triangular apertural opening with height of 27 mm, spanning prosiphuncular part of body chamber height. Horn-like projection forms rounded triangle with depression that opens towards prosiphuncular side (Pl. 1, figs 10, 14). Length of contracted part of body chamber 10 mm. Shell thickness close to peristome 1.1 mm. Shell smooth. Septal distance of last two chambers 4 mm. Septal curvature shallow concave with deepest point of curvature between growth axis and prosiphuncular conch margin. Septa oblique, slope towards apex on prosiphuncular side. Sutures nearly straight, form faint lateral lobe. Siphuncle positioned *c.* 3 mm from conch margin where conch cross-section is 35 mm. Septal perforation compressed ovate, width 5.7 mm, height 7.1 mm.

Remarks. Cerovoceras fatimi differs from *C. brevidomus* in having a longer body chamber with a circular cross-section and a smaller adult size.

Occurrence. Early Eifelian; Morocco.

Cerovoceras brevidomus sp. nov.
Plate 1, figures 12–13; Text-figure 12D–F

Derivation of name. Latin, *breve*, short, and *domus*, house, referring to the short body chamber of the species.

Holotype. MB.C.9620 (Pl. 1, figs 12–13).

Type locality and horizon. Filon Douze section, bed PI, early *O. costatus* conodont Biozone, middle Eifelian.

Other material. Known only from the holotype.

Diagnosis. Cerovoceras with adult size of *c.* 51 mm in conch height; slightly depressed cross-section; shell surface smooth; length of mature body chamber *c.* 39 mm; aperture broadly rounded-triangular, flattened side on antisiphuncular side of conch, height *c.* 22 mm, width 34 mm; broadened subtriangular projection of body chamber on antisiphuncular side 20 mm adoral of last septum; adjacent, contracted part of body chamber *c.* 10 mm long; septa slightly oblique, sloping adapically on prosiphuncular side.

Description. Holotype is fragment of body chamber and parts of latest chamber of phragmocone. Conch height at base of body chamber 48 mm, conch width *c.* 46 mm. Body chamber length from latest septum to peristome *c.* 39 mm. Maximum gibbosity of body chamber 20 mm, occurs adoral of base of body chamber. At this position on antisiphuncular side conch projection occurs and just adoral of projection; body chamber constricted forming broadened, rounded, triangular apertural opening 21 mm in height, spanning prosiphuncular part of body chamber height. Horn-like projection forms broadened rounded triangle with slight depression that opens towards prosiphuncular side. Length of contracted part of body chamber 12 mm. Septal distance between last two septa 6 mm. Septal curvature shallow concave with deepest point between growth axis and prosiphuncular conch margin. Septa oblique, slope towards the apex on prosiphuncular side. Sutures nearly straight, form faint lateral lobe. Siphuncle position and shape of septal perforation not visible on specimen.

Remarks. Cerovoceras brevidomus differs from *C. fatimi* in having a larger adult size, a shorter body chamber, a slightly depressed cross-section, and a laterally elongated and less conspicuous body chamber horn.

Occurrence. Middle Eifelian; Morocco.

Family JOVELLANIIDAE Foord, 1888

Genus JOVELLANIA Bayle, *in* Bayle and Zeiller 1878

Type species. Orthoceratites buchi Verneuil, 1850, from the Siegenian, Early Devonian of Néhou, Manche, France.

Diagnosis. Straight or slightly cyrtoconic, slowly widening conch with faint undulation; cross-section nearly circular or slightly depressed with flattened ventrolateral sides causing slightly angular prosiphuncular side; sutures directly transverse; septa spaced very closely; siphuncle positioned between centre and conch margin on convex side of growth axis, closer to conch margin than to growth axis; siphuncular segments expanded within chambers; endosiphuncular deposits consist of longitudinal lamellae.

Remarks. The diagnosis above is a synthesis of generic diagnoses provided by Foerste (1926, p. 306) and Zhuravleva (1974, p. 44). No information on the septal neck shape was given previously. Serpagli and Gnoli (1977, p. 95) described cyrtochoanitic septal necks in *Jovellania*, whereas in the specimens from Filon Douze they are recumbent. However, because no information on the intra-generic variability of this character is known, the diagnostic value of the septal neck remains unclear.

Occurrence. Lochkovian?; France, Germany, Morocco and Tibet.

Jovellania cheirae sp. nov.
Plate 1, figures 8, 11; Plate 2, figures 1, 3–4; Plate 3, figures 1–4

Derivation of name. In honour of Cheira, wife of Filon Douze miner Lahcan Caraoui.

Holotype. MB.C.9603 (Pl. 3, figs 1–3).

Type locality and horizon. Filon Douze section, bed OJ-II, Lochkovian.

Other material. 8 fragments, MB.C.9604–9611, 9621, from bed OJ-I, Lochkovian.

Diagnosis. Jovellania with slightly cyrtoconic, slowly enlarging conchs; shell surface smooth, only faintly undulating; cross-section slightly depressed with flattened ventrolateral sides forming slightly angulate prosiphuncular side; conch slightly bilaterally asymmetric; sutures form shallow lateral saddles and shallow, slightly angular lobe on prosiphuncular side; nine septa in length similar to conch diameter; siphuncle close to conch margin; septal necks recumbent; septal perforation elongate oval in cross-section.

Description. Conch surface smooth with faint undulations, nearly straight conch in adult growth stages but slightly cyrtoconic in juvenile stages (Pl. 3, fig. 4). One adult specimen (MB.C.9621) is known, a fragment of parts of phragmocone and body chamber. Body chamber 59 mm long, 51 mm wide, nearly tubular shape, faint longitudinal striae at base. Most adoral chambers of specimen crowded. Holotype 56 mm long, 35 mm (smallest)–41 mm (largest) in diameter, nearly circular in cross-section. Ventrolateral sides of conch slightly elongate giving cross-section slightly trapezoidal shape. Entire cross-section a little bilaterally asymmetrical (Pl. 3, fig. 3). Septal perforation slightly off bilateral axis. Fragments show some variation in shape of cross-section. Ratio of conch height to width 0.9–1. Septal distance in holotype varies between 3.2 and 4.6 mm. Sutures describe wide saddle on lateral sides and narrower lobe on prosiphuncular side. Septal perforation 2 mm from conch margin; conch ovate with height 7.2 mm and width 4.9 mm at most adoral septum of holotype (0.18 of conch diameter). In specimen MB.C.9604, height of siphuncle 5.8 mm (at conch

diameter 40 mm), septal distance 3.1 mm and septal perforation 2.5 mm from conch margin. In this specimen connecting ring is preserved, showing siphuncular expansion between septa of *c.* 1 mm. Septal necks recumbent; endosiphuncular deposits composed of longitudinal lamellae.

Remarks. This species differs from *J. buchi* in having a nearly smooth shell with only very faint undulation and lower expansion rate; it differs from all other species of the genus in having a ratio of conch height to width of 0.9–1, and a nearly smooth shell.

Genus BOHEMOJOVELLANIA Manda, 2001

Type species. Bohemojovellania bouskai Manda, 2001, from the Praha Formation, Pragian, Czech Republic.

Diagnosis (after Manda 2001, p. 274). Cyrtoconic jovellanids with smooth, sinistral, bilaterally asymmetric conch; siphuncle on convex side of conch curvature; cross-section elliptically depressed; *c.* 10 septa per length similar to conch width; septa moderately concave, sinistrally oblique to horizontal axis, with maximum depth of curvature between dorsoventral axis and convex conch margin; siphuncle shifted off dorsoventral axis; siphuncular segments one-seventh of conch width in diameter.

Remarks. The otherwise similar *Anonymoceras* Zhuravleva, 1974 differs in having a bilaterally symmetrical conch.

Occurrence. Lochkovian–Pragian; Czech Republic and Morocco.

Bohemojovellania bouskai Manda, 2001
Plate 3, figure 9

*2001   *Bohemojovellania bouskai* Manda, pp. 275, 280, pl. 1, figs 13–15, text-fig. 6.

Material. 2 fragments, MB.C.9616–17, from bed OJ-I, Lochkovian, Filon Douze section.

Diagnosis (after Manda 2001, p. 275). Slightly cyrtoconic *Bohemojovellania* with depressed elliptical cross-section, ratio of conch height to width 0.8; cross-section depressed rounded-trapezoidal with left side of conch slightly more rounded at position of siphuncular perforation, which is close to margin on convex side of growth axis, slightly shifted from dorsoventral axis to left; septa oblique, slightly sloping (8 degrees) in adoral direction on concave side of curvature of growth axis and on right side; septal curvature shallow with deepest point of curvature at mid-length between cross-section centre of gravity and position of siphuncle; sutures form shallow, wide saddles half-way between dorsoventral axis and right conch margin at convex and concave side of growth axis; ten septa per length similar to conch height; septal perforation circular in cross-section and *c.* 0.8 of conch diameter.

Description. Conch smooth, very slightly cyrtoconic. MB.C.9616 is 11 mm long, 21 mm wide and 18 mm high, displaying four chambers. MB.C.9617 is 9 mm long, 34 mm wide and 27 mm high, displaying three chambers. Considering cross-section subdivided into 360 degrees, axis of shell curvature along 0–180 degrees (= dorsoventral axis), with position 0 degrees directed to concave side of curvature of growth axis (=dorsal). Siphuncle in both fragments is close to conch margin at 190 degrees. Septa slightly oblique along right–left axis of 90–270 degrees sloping in apical direction towards left side. Deepest point of septal curvature close to centre of gravity of cross-section. Sutures form wide saddles half-way between dorsoventral axis and right conch margin. Septal perforation circular, 1.3 mm wide in MB.C.9617, 2.6 mm in MB.C.9616.

Occurrence. Lochkovian; Morocco.

Bohemojovellania adrae sp. nov.
Plate 2, figures 2, 7; Plate 3, figures 5–8

Derivation of name. In honour of my Moroccan friend Adrae.

Holotype. MB.C.9597 (Pl. 3, figs 5–8).

Type locality and horizon. Filon Douze section, bed OJ-I, Lochkovian.

Other material. 4 fragments, MB.C.9598–9601, from the type locality and horizon.

Diagnosis. Bohemojovellania with depressed elliptical cross-section, ratio of conch height to width 0.8; elongate,

flattened, ventrolateral conch sides and slightly angulate prosiphuncular conch side; cross-section slightly bilaterally asymmetrical with siphuncle slightly shifted off axis of symmetry; sutures slightly oblique, sloping adorally on prosiphuncular side; septa have strong curvature; eight septa per length similar to conch height; siphuncle close to conch margin; septal perforation circular in cross-section.

Description. Conch slightly cyrtoconic, surface smooth with faint undulations. Holotype 37 mm long, 29–32 mm wide, 23–26 mm high; cross-section broadly elliptical with flattened antisiphuncular side. Septal perforation shifted slightly off the axis of symmetry. Ratio of conch height to width 0.76–0.81, and in some specimens, antisiphuncular side less flattened than in holotype, making cross-section nearly perfectly elliptical. Septal distance of holotype varies between 1.7 and 3.0 mm. Sutures oblique towards growth axis, sloping towards the aperture on prosiphuncular side (Pl. 3, fig. 5), describing wide saddle on prosiphuncular side and wide lobe on antisiphuncular side. Septa strongly concave with deepest point of curvature half-way between centre and margin of antisiphuncular side. Septal perforation marginal on concave side of growth axis, circular, diameter 2.6 mm in holotype. Diameter of septal perforation in holotype 0.11 of conch height. Septal necks recumbent and endosiphuncular deposits composed of longitudinal lamellae (Pl. 2, fig. 7).

Remarks. This species differs from all other species of *Bohemojovellania* in having an only slightly asymmetrical conch. *Anonymoceras acceptatum* Zhuravleva, 2000, differs in having a less angular venter in cross-section and a symmetrical conch.

Bohemojovellania obliquum sp. nov.
Plate 3, figures 10–12

Derivation of name. Latin, *obliquus*, oblique.

Holotype. MB.C.9636 (Pl. 3, figs 10–12).

EXPLANATION OF PLATE 2

Median sections of Oncocerida from the Filon Douze section.

Figs 1, 3–4. *Jovellania cheirae* sp. nov., MB.C.9604; bed OJ-I, Lochkovian. 1, entire fragment. 3, detail showing conspicuous shape of connecting ring, recumbent septal necks and lamellar endosiphuncular deposits. 4, detail showing septal neck.

Figs 2, 7. *Bohemojovellania adrae* sp. nov., MB.C.9601; bed OJ-I, Lochkovian. 2, entire specimen. 7, detail showing conspicuous shape of connecting ring, cyrtochoanitic septal necks and lamellar endosiphuncular deposits.

Figs 5, 8. *Ankyloceras* sp., MB.C.9626; bed OSP-II, Přídolí. 5, entire fragment. 8, detail showing conspicuous shape of connecting ring; note the asymmetric shape of septal necks and annular endosiphuncular deposits.

Figs 6, 9. *Ventrobalashovia zhuravlevae* gen. et sp. nov., MB.C.9637, fragment of holotype; bed KMO-III, Pragian. 6, entire fragment. 9, detail showing slightly expanded siphuncular tube, orthochoanitic septal necks and lamellar endosiphuncular deposits.

Figs 10–11. *Tafilaltoceras adgoi* gen. et sp. nov., MB.C.9634, holotype; bed KMO-I, Pragian. 10, entire specimen. 11, detail showing nearly tubular siphuncular tube, orthochoanitic septal necks and lamellar endosiphuncular deposits.

Scale bars represent 1 mm; figures without a scale bar are × 1.

PLATE 2

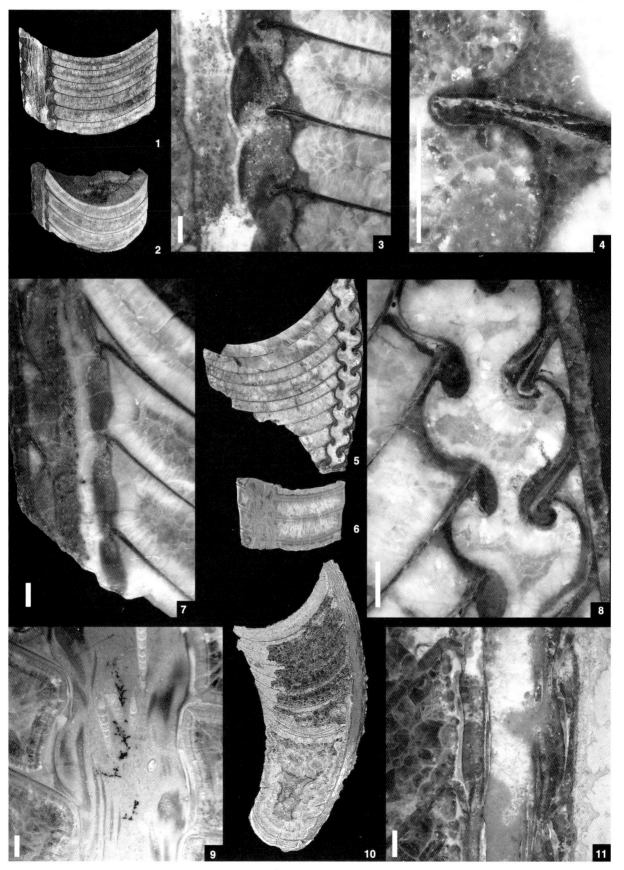

Type locality and horizon. Filon Douze section, bed OJ-I, Lochkovian.

Other material. 2 fragments, MB.C.9602, 9691, from the type locality and horizon.

Diagnosis. Slightly cyrtoconic *Bohemojovellania* with depressed elliptical cross-section, ratio of conch height to width 0.8; left side of conch slightly more rounded at position of siphuncular perforation, which is close to conch margin on convex side of conch growth axis, strongly shifted from dorsoventral axis to left; septa oblique, sloping in adoral direction on concave side of curvature of growth axis and on right side; septal curvature shallow, with deepest point of curvature at mid-length between cross-section centre of gravity and position of siphuncle; sutures form shallow saddles half-way between dorsoventral axis and right conch margin on convex and concave sides of growth axis; ten septa per length similar to conch height; septal perforation circular in cross-section and *c.* 0.8 of conch diameter.

Description. Conch smooth, very slightly cyrtoconic with angle of expansion of 5 degrees. Only fragments of phragmocone known. Holotype 37 mm long, 28–32 mm wide, 22–29 mm high. Considering cross-section subdivided into 360 degrees, axis of shell curvature along 0/180 degrees (= dorsoventral axis) and 0 degrees directed towards centre of growth curvature (= direction of concave side of curvature of growth axis). Siphuncle in holotype 2 mm from conch margin at position of 250 degrees (Pl. 3, fig. 12). Septa oblique along right/left axis 90–270 degrees, sloping in apical direction on left side of (Pl. 3, fig. 10). Deepest point of septal curvature half-way between centre of gravity of cross-section and position of siphuncle. Sutures form wide saddles half-way between dorsoventral axis and right conch margin. Septal perforation in holotype 2.6 mm wide and circular. Septal perforation of MB.C.9691 2 mm from conch margin at 225 degrees, with circular cross-section of 4.5 mm; this specimen is a fragment with five chambers with septal distance of 3.5 mm, conch width 52 mm, conch height 39 mm.

Remarks. This species differs from *B. bouskai* in having a more oblique, eccentric siphuncle.

Family KAROCERATIDAE Teichert, 1939

Genus ANKYLOCERAS Zhuravleva, 1974

Type species. Cyrtoceras nesnayamiense Foerste, 1925, from the Lower Devonian of Novaya Zemlya, Russia.

Diagnosis (after Zhuravleva 1974, p. 67). Cyrtocone karoceratids with compressed to slightly depressed cross-section, ornamented with faint growth lines displaying distinct lobe on convex and sometimes on concave side of growth axis; cross-section elliptical and more narrowly vaulted on convex side of growth axis curvature; suture line forms broad lateral lobes; siphuncle marginal on convex side of conch growth axis, scalariform; siphuncular segments strongly expanded towards conch margin at adapical part of each segment, and towards conch centre at adoral part of each segment, at expanded portions touching surface of septa; septal necks loxochoanitic on side directed towards conch margin, short, recumbent on side directed towards growth axis.

Occurrence. Early Devonian; Japan, Morocco, Novaya Zemlya and Russia.

Ankyloceras sp.
Plate 2, figures 5, 8; Plate 3, figures 13–14

Material. 1 specimen, MB.C.9626, from bed OSP-II, Přídolí, Filon Douze section.

Description. Fragment of phragmocone, length *c.* 42 mm. Only convex half of cyrtoconic conch preserved. Cross-section appears to be circular or slightly depressed with diameter at adoral end 28 mm. Siphuncle marginal on convex side of conch curvature. Conch surface smooth with slightly irregular transverse undulations. Septal spacing slightly irregular with septal distances of 1.5–2.6 mm. Septal perforation depressed oval, 1.8 mm in cross-section. Septal necks cyrtochoanitic on side of siphuncle directed towards conch margin and strongly recumbent on side directed towards conch centre. Siphuncular segments strongly expanded. Connecting ring forms rounded-convex segments that touch adapical surface of septa on side directed towards growth axis of conch and touch adoral septal surface on side directed towards conch margin. Endosiphuncular deposits form adapically elongated bullettes at position of septal perforation.

Remarks. In contrast to the type of the genus, this specimen has ventrally directed septal necks that are cyrtochoanitic, not loxochoanitic. However, as there is no information of the variability of this character within the genus, I consider the difference small enough to allow assignation to *Ankyloceras*. Because the single specimen reveals no information on the general conch shape, a specific designation is impossible. The specimen may be conspecific with *Ankyloceras sardoum* (Serpagli and Gnoli, 1977) comb. nov. but its curvature is stronger, and the segments of the siphuncle are more expanded in the latter.

Genus VENTROBALASHOVIA gen. nov.

Type species. Ventrobalashovia zhuravlevae sp. nov., from the Pragian of the Filon Douze section.

Derivation of name. Latin, *venter*, belly, referring to the ventral siphuncle of this genus, which is otherwise similar to *Balashovia* Zhuravleva, 1974.

Other species included. Tripleuroceras altajense Zhuravleva, 1979 (species combined with *Ventrobalashovia* below: see 'List of new combinations').

Diagnosis. Slightly cyrtoconic oncocerids with undulating shell; cross-section with rounded margin on convex side of growth axis and flattened margin on concave side; siphuncle positioned on convex side of conch curvature, slightly removed from conch margin; *c.* 8 septa per length similar to conch width; septal curvature very shallow; siphuncular diameter 0.2–0.3 of conch height; siphuncle only slightly expanded within chambers; septal necks orthochoanitic to loxochoanitic, relatively long; endosiphuncular deposits as longitudinal lamellae.

Remarks. This genus is similar to typical jovellanids in the general conch shape. However, *Jovellania* has cyrtochoanitic or recumbent septal necks. By contrast, the long orthochoanitic to loxochoanitic septal necks of *Ventrobalashovia* are similar to *Balashovia* Zhuravleva, 1974, which was assigned to the Tripleuorthoceratidae because it has a siphuncle on the concave side of conch curvature and a flattened venter. *Ventrobalashovia* has a siphuncle on the convex side of shell curvature and a flattened dorsum, and thus differs from genera assigned to both the Jovellaniidae and the Tripleurorthoceratidae. This relationship problem can only be solved when the internal characters of *Tripleuroceras* Hyatt, 1884 are better known and the Tripleuroceratidae are revised.

Occurrence. Pragian–Eifelian; Altai Mountains (Russia) and Morocco.

Ventrobalashovia zhuravlevae sp. nov.
Plate 2, figures 6, 9; Plate 3, figures 15–17

Derivation of name. In honour of F. A. Zhuravleva, Moscow, for her lifelong research on nautiloids.

Holotype. MB.C.9637 (Pl. 2, figs 6, 9; Pl. 3, figs 15–17).

Type locality and horizon. Filon Douze section, bed KMO-III, Pragian.

Other material. Known only from the holotype.

Diagnosis. Ventrobalashovia with obliquely undulating shell; two undulations per distance similar to length of two chambers; undulations slope towards aperture on prosiphuncular side of conch; cross-section with rounded

margin on convex side of growth axis, and flattened on concave side of this axis, with height to width ratio of 0.8; siphuncle at convex side of conch curvature slightly removed from margin; about eight septa per distance similar to conch width; septal curvature very shallow; siphuncular diameter 0.2 of conch height; siphuncle only slightly expanded within chambers; septal necks orthochoanitic to loxochoanitic, relatively long; endosiphuncular deposits as longitudinal lamellae.

Description. Conch slightly cyrtoconic with angle of expansion of 7 degrees. Holotype 58 mm long, 34–50 mm wide, 32–40 mm high. Conch undulating with distance between two undulations 11 mm; these slope in adoral direction on prosiphuncular side. Cross-section nearly elliptical, flattened on concave side of growth axis (Pl. 3, fig. 17). Septa straight with shallow concavity. Sutures form shallow, narrow lobe on prosiphuncular side. Septal perforation 5 mm from conch margin on convex side of conch curvature. Septal necks orthochoanitic or loxochoanitic, 2.5 mm long. Siphuncular segments only slightly expanded. Difference between diameter of septal perforation and maximum diameter of siphuncle 1.8 mm. Endosiphuncular deposits consist of longitudinal lamellae.

Remarks. This species differs from *Tripleuroceras altajiense* Zhuravleva, 1979 in having a wider cross-section, a more eccentric, narrower siphuncle and a higher expansion rate.

Family NOTHOCERATIDAE Fischer, 1882

Genus MUTOBLAKEOCERAS gen. nov.

Type species. Mutoblakeoceras inconstans sp. nov., from the Eifelian of the Filon Douze section.

Derivation of name. Latin, *muto*, to change, with reference to the ontogenetically changing cross-section shape of the genus, which is otherwise similar to *Blakeoceras* Foerste, 1926.

Other species included. None.

Diagnosis. Slightly cyrtoconic nothoceratids with angle of expansion of 18 degrees; cross-section compressed adapically, depressed in later growth stages; septa straight, transverse and with shallow concavity; about eight septa per distance similar to cross-section; septal perforations *c.* 0.1 of conch cross-section; marginal siphuncle at convex side of shell curvature; siphuncular segments tubular with thick outer layer of connecting rings, thickest in vicinity of septal perforation, giving impression of concave segments; septal necks suborthochoanitic; endosiphuncular deposits as continuous longitudinal lamellae.

Remarks. The typical concave appearance of the siphuncular segments in the type qualifies the genus as a nothoceratid. This large species differs from all other nothoceratids in having a cross-section that changes during ontogeny from compressed to depressed. The genus differs from the otherwise similar *Blakeoceras* in having a less curved conch and a compressed to depressed cross-section.

Occurrence. Eifelian; Morocco.

Mutoblakeoceras inconstans sp. nov.
Plate 4, figures 2–3, 5; Plate 5, figure 1

Derivation of name. Latin, *inconstans*, inconsistent, with reference to the ontogenetic change in the cross-section of this species.

Type locality and horizon. Filon Douze section, bed F, *P. partitus* conodont Biozone, early Eifelian.

Other material. Known from the holotype only.

Diagnosis. As for genus, by monotypy.

Description. Specimen is fragment of very slightly cyrtoconic phragmocone 92 mm long; conch width 34–64 mm, height 39–56 mm. Thus, conch adapically compressed and adorally depressed in cross-section with angle of expansion of *c.* 18 degrees (Pl. 4, figs 2, 5). Siphuncle marginal at convex side of conch curvature. Fragment has 11 chambers. Septa slightly oblique, sloping in adoral direction on prosiphuncular side. Septal spacing slightly irregular with septal distances between 5 mm (adapically) and 11 mm (adorally). Diameter of septal perforation 6.4 mm at adoral end.

Genus TAFILALTOCERAS gen. nov.

Type species. *Tafilaltoceras adgoi* sp. nov., from the Pragian of the Filon Douze section.

Derivation of the name. After Tafilalt, with reference to the type locality.

Other species included. None.

Diagnosis. Gradually expanding cyrtoconic nothoceratids with angle of expansion of 12 degrees; cross-section subcircular, slightly depressed; septa straight, transverse and with shallow concavity; about nine septa per distance similar to cross-section; septal perforations *c.* 0.16 of conch cross-section; marginal siphuncle at convex side of shell curvature; siphuncular diameter *c.* 0.1 of cross-section diameter; siphuncular segments tubular with thick outer layer of connecting rings, thickest in vicinity of septal perforation, giving impression of concave segments; septal necks suborthochoanitic; endosiphuncular deposits as continuous longitudinal lamellae.

Remarks. Although the ornamentation of *Tafilaltoceras* is not known in detail because the holotype is heavily overgrown by epizoans, it is apparently smooth and clearly without undulations. It differs from the otherwise similar *Blakeoceras* Foerste, 1926, in having a slightly depressed cross-section, from *Conostichoceras* Foerste, 1926, in having a gradually widening, more elongate conch, and from *Turneroceras* Foerste, 1926, in lacking a flattened venter.

A diagnostic character of the Nothoceratidae is the 'concavosiphonate siphuncle' (Sweet 1964a, p. K305). In *Tafilaltoceras* the siphuncle also appears to be concavosiphonate. However, the better preserved areas in the type reveal that the siphuncular segments are tubular and the apparent vacuosiphonate (=concavosiphonate) shape of the siphuncle is an effect of the very conspicuous, thick outer layer of the connecting ring, thickest in the vicinity of the septal perforation (Pl. 2, fig. 11). It is highly probable that this is also the case in other nothoceratidans.

Occurrence. Pragian; Morocco.

PLATE 3

KRÖGER, nautiloids

Tafilaltoceras adgoi sp. nov.
Plate 2, figures 10–11

Derivation of name. In honour of Adgo, youngest son of Filon Douze miner Lahcan Caraoui.

Holotype. MB.C.9634 (Pl. 2, figs 10–11).

Type locality and horizon. Filon Douze section, bed KMO-I, Pragian.

Other material. Known only from the holotype.

Diagnosis. As for the genus.

Description. Fragment of phragmocone 77 mm long, conch height 18–34 mm, adoral width 37 mm. Cross-section slightly depressed. Conch surface not known but no undulations or evidence for ornamentation visible in cross-section. Septal distance at adoral end of specimen 4.5 mm. Septal curvature shallow concave. Siphuncle positioned at margin on convex side of conch curvature. Septal perforation at most adoral septum 5.5 mm, where thickness of siphuncle is 4.3 mm. Siphuncular segments nearly tubular. Connecting ring displays thick outer layer that expands over the adapical surfaces of suborthochoanitic septal necks, forming wedge at adapical surface of septal neck and at adoral surface of septal perforation.

Indet. Nothoceratidae sp. A.
Plate 4, figure 1; Plate 5, figure 2

Material. 1 specimen, MB.C.9635, from bed PI, early *O. costatus* conodont Biozone, middle Eifelian; Filon Douze section.

Description. Specimen is fragment of phragmocone 83 mm long with smooth shell. Cross-section slightly depressed; maximum conch width 67 mm, conch height 71 mm. Angle of expansion *c.* 21 degrees. Siphuncle marginal at convex side of conch curvature. Twelve chambers preserved. Septal spacing slightly irregular with septal distances of *c.* 5 mm. Septal perforation has diameter of 10 mm at adoral end. Septal necks orthochoanitic. Siphuncular segments appear concave. Endosiphuncular deposits occur adapically as longitudinal lamellae.

Remarks. This specimen differs from *Mutoblakeoceras inconstans* in having a larger angle of expansion and a depressed cross-section. A depressed cyrtoconic phragmocone is characteristic of *Turnoceras* and *Constichoceras*, which differ mainly in the shape of the adult body chamber. Because the body chamber is not preserved in the Filon Douze fragment a generic or specific determination cannot be made.

Family ONCOCERATIDAE Hyatt, 1884

Genus ORTHORIZOCERAS gen. nov.

Type species. Orthorizoceras desertum sp. nov., from the upper Přídolí of the Filon Douze section.

Derivation of the name. Greek, *orthos*, straight, referring to the straight conch.

Other species included. Metarizoceras sinkovense Balashov, *in* Balashov and Kiselev 1968 (species combined with *Orthorizoceras* below: see 'List of new combinations').

Diagnosis. Straight, rapidly widening, breviconic oncocerids with slightly compressed to slightly depressed cross-section; about five septa at distance similar to conch cross-section; septa slightly oblique, sloping towards apex on prosiphuncular side; sutures form shallow lateral lobe; siphuncle close to conch margin; siphuncular segments expanded within chambers; septal necks cyrtochoanitic; endosiphuncular and cameral deposits not known.

Remarks. Orthorizoceras is similar to *Rizoceras* and *Metarizoceras* Foerste, 1930, but differs in having a straight conch that has its largest diameter adapical of the base of the body chamber. The conch of *Rizoceras* gradually widens, and that of *Metarizoceras* has its largest diameter at the base of the body chamber. The septal spacing of *Orthorizoceras* is also much greater than that of these two genera.

Occurrence. Přídolí; Ukraine and Morocco.

Orthorizoceras desertum sp. nov.
Plate 4, figure 4; Plate 9, figure 8

Derivation of name. Latin, *deserta*, desert, referring to the type locality, which is at the northern margin of the Sahara.

Holotype. MB.C.9633 (Pl. 4, fig. 4; Pl. 9, fig. 8).

Type locality and horizon. Filon Douze section, bed OSP-II, Přídolían.

Other material. Known only from the holotype.

Diagnosis. Orthorizoceras with angle of expansion of *c.* 20 degrees; maximum cross-section diameter adapical of body chamber *c.* 30 mm; cross-section slightly depressed, height to width ratio 0.9; septa slightly oblique, sloping towards apex on prosiphuncular side; sutures form shallow lateral lobe; siphuncle close to conch margin; siphuncular segments expanded within chambers; septal necks cyrtochoanitic; endosiphuncular and cameral deposits not known.

Description. Fragment of phragmocone 36 mm long, conch 14–29 mm high, width adapically 16 mm. Maximum diameter of cross-section at position of second youngest chamber. Outer shell not preserved. Septal crowding of most adoral septa indicates maturity of specimen. Septal distance at adoral end of specimen 3 mm, at apical end 4 mm. Maximum septal distance 5 mm. Septal curvature shallowly concave. Septal perforation positioned 1.5 mm from conch margin at most apical septum, *c.* 0.1 of cross-section. Siphuncular segments expanded within chambers. Siphuncular diameter 3 mm at apical end of specimen.

Remarks. This species differs from *Metarizoceras sinkovense* Balashov, *in* Balashov and Kiselev 1968 in having a smaller adult size and wider septal spacing.

Order NAUTILIDA Agassiz, 1847
Family CENTROCERATIDAE Hyatt, 1900

Genus CENTROCERAS Hyatt, 1884

Type species. *Goniatites marcellensis* Vanuxem, 1842, from the Cherry Valley limestone, Eifelian, of central New York, USA.

Diagnosis (after Kummel *et al.* 1964, p. K434). Tarphyceraconic nautiloid consisting of few, rapidly expanding volutions; umbilical window wide; slightly impressed zone near base of mature living chamber; whorl sections tetragonal, with sharp ventral and umbilical shoulders; dorsum broad, convex in young, nearly straight in mature, portions; lateral areas oblique, convergent towards narrow, scarcely convex venter; whorls compressed; sutures form shallow ventral lobe and broad lateral lobe with subacute saddles on shoulders, dorsal portion nearly transverse except in mature portion where broad, shallow, lobe occurs; siphuncle tubular, positioned near venter; living chamber comprises half a volution in length; ornamented with alternating striae and lirae which are more or less fasciculate; ventral shoulders of early volutions with small nodes.

Remarks. The record of *Centroceras* sp. from the Emsian of Filon Douze is the oldest of the genus and of the Nautilida in general.

Occurrence. Zlíchovian–Givetian; Armenia, Germany, Morocco and North America.

Centroceras sp.
Plate 4, figures 11–13

Material. 1 specimen, MB.C.9628, from the top of bed KMO-IV, Zlíchovian, Emsian, Filon Douze section.

Description. Specimen displays eight chambers of intermediate growth stages of conch belonging to most adoral part of phragmocone. Additionally, cast of outer shell of innermost whorls is preserved. Diameter at position of youngest preserved septum 62 mm. Maximum whorl height of preserved chambers *c.* 32 mm, minimum preserved whorl height *c.* 25 mm. Length of curvature of conch part 52 mm. Septa at distances of 5–9 mm. Most adoral four preserved chambers crowded, giving evidence for maturity of specimen. Septa deeply concave, oblique towards radial axis, sloping towards apex on convex conch side. Dorsum broad, with straight margin in adoral part, maximum width 46 mm, convex in adapical part. Lateral sides rounded, ornamented with faint, irregularly spaced, transverse undulations and lirae that form conspicuous lobe at venter (Pl. 4, fig. 13). Convex side of shell not preserved. Septal perforation not visible, but it cannot be central or between centre and dorsum.

Remarks. This specimen shows diagnostic features of *Centroceras* such as the general shell shape, shape of septa and sutures, and ornamentation. However, the conch venter is only poorly preserved. Because this character is important for assignment to a species, such a course of action for this single fragment is impossible, and the exact position of the siphuncle is unknown.

Order ACTINOCERIDA Teichert, 1933
Family ORMOCERATIDAE Saemann, 1852

Genus ORMOCERAS Stokes, 1840

Type species. *Orthoceras bayfieldi* Stokes, 1838, from the Niagaran (Wenlock) of Drummond Island, Lake Huron, Canada.

Diagnosis (after Teichert 1964, p. K211). Conch straight with nearly circular cross-section; siphuncle subcentral with almost globular segments; septal necks cyrtochoanitic with short brims; annulosiphuncular deposits with simple endosiphuncular canal system and few radial canals; cameral deposits common.

Occurrence. Middle Ordovician–Early Devonian; world-wide.

Ormoceras sp.
Plate 5, figure 3; Text-figure 13B

Material. 1 specimen, MB.C.9639, from bed KMO-I, Pragian, Filon Douze section.

Description. Fragment of six chambers of phragmocone, straight, length 29 mm, cross-section diameter adorally *c.* 18 mm. Because of fragmentary character of specimen, angle of expansion cannot be determined. Minimum distance between septa 4.5 mm (at adoral end of fragment), maximum distance

5.2 mm. Septal concavity shallow with 2.8 mm curvature depth at adapical septum. Siphuncle eccentric. Septal perforation diameter 3.1 mm where cross-section diameter is 17 mm. Diameter of siphuncle *c.* 5 mm. Siphuncular segments subglobular, widely expanded within chambers with height to width ratio of 0.8. Septal necks cyrtochoanitic. Brims of septal necks display conspicuous sharp bend at distal one-third of necks. Annulosiphuncular deposits thicker on ventral side of siphuncle. Cameral deposits occur on prosiphuncular side of conch.

Remarks. The fragment is assigned to *Ormoceras* because of the presence of cyrtochoanitic septal necks that display a conspicuous sharp bend at their brims. The widely expanded siphuncular segments are subglobular in longitudinal section. However, the fragmentary nature of the single specimen does not allow a specific determination.

Genus DEIROCERAS Hyatt, 1884

Type species. *Orthoceras python* Billings, 1857, from the Cobourg Beds, Ottawa Formation, Caradoc, Late Ordovician, of Ottawa, Ontario, Canada.

Diagnosis (after Teichert 1964, p. K212). Slender, gradually expanding, orthoconic longicones; conch cross-section nearly circular; septal spacing wide; septal necks suborthochoanitic; siphuncle between conch margin and centre; siphuncular segments longer than wide, ovate; annulosiphonate deposits with simple, straight radial canals perpendicular or slightly oblique to narrow central canal; with episeptal and hyposeptal deposits.

Occurrence. Middle Ordovician–Early Devonian; world-wide.

Deiroceras hollardi sp. nov.
Plate 5, figures 7–8; Plate 16, figure 8; Text-figure 13A

Derivation of name. In honour of Henri Hollard who was a pioneering researcher on the geology and stratigraphy of Morocco.

Holotype. MB.C.9630 (Pl. 5, figs 7–8; Pl. 16, fig. 8; Text-fig. 13A).

Type locality and horizon. Filon Douze section, *Deiroceras* Limestone, Zlíchovian, Emsian.

Other material. 3 specimens, MB.C.9640, 9768.1–2, from the type locality and horizon.

Diagnosis. *Deiroceras* with angle of expansion of 2 degrees, ornamented with fine transverse lirae; conch cross-section nearly circular; septal spacing wide, *c.* 2 septa per distance similar to conch diameter; diameter of septal perforation *c.* 0.27 of conch cross-section; septal necks suborthochoanitic; siphuncle subcentral; siphuncular segments elongate, ovate; annulosiphonate deposits with simple, straight radial canals closer to adapical surface of each septum than to adoral surface and perpendicular or slightly oblique to narrow central canal; with episeptal and hyposeptal deposits.

Description. Conch with very low angle of expansion (*c.* 2 degrees). Shell surface ornamented with fine transverse lirae; about ten lirae per 2 mm (Pl. 16, fig. 8). Septal distance in holotype 19 mm at conch diameter of 41 mm, in MB.C.9640, 14 mm at conch diameter of 30 mm. Septa have deep curvature and are slightly conical in shape. Septal perforation subcentral with diameter of 11 mm in holotype (0.27 of conch diameter), 6.5 mm in MB.C.9640 (0.22 of conch diameter). Siphuncular segments elongate-ovate. Diameter of siphuncular segment at mid-length of holotype 17.7 mm (0.43 of diameter), in MB.C.9640 11.8 mm (0.39 of conch diameter). Endosiphuncular deposits consist of massive annuli, slightly elongated in adoral direction. Contact surface between two successive annuli slightly oblique towards growth axis and central canal. In holotype, adapical contact surface of endosiphuncular annuli occurs 7.4 mm from septal perforation, adoral contact surface 12 mm from it. Central canal appears to be off-centre of siphuncle. However, in MB.C.9640 median section displays centrally positioned central canal. On prosiphonal side of conch episeptal and hyposeptal deposits occur.

EXPLANATION OF PLATE 4

Fig. 1. Indet. Nothoceratidae sp. A., MB.C.9635; bed PI, early *O. costatus* conodont Biozone, middle Eifelian; dorsal view.

Figs 2–3, 5. *Mutoblakeoceras inconstans* gen. et sp. nov., MB.C.9584; bed PI, early *O. costatus* conodont Biozone, middle Eifelian. 2, adapical view. 3, lateral view. 5, adoral view; note the depressed adoral cross-section and the compressed adapical cross-section.

Fig. 4. *Orthorizoceras desertum* gen. et sp. nov., MB.C.9633, holotype; bed OSP-II, Přídolí; lateral view.

Figs 6–7. *Subormoceras* sp., MB.C.9674; bed OSP-I, Přídolí. 6, lateral view. 7, adapical view; note the clear asymmetric position of siphuncle.

Figs 8–9. *Orthocycloceras?* *fluminense* (Meneghini 1857), MB.C.9612; bed OSP-I, Přídolí. 8, ventral view. 9, lateral view.

Fig. 10. *Subormoceras rissaniense* gen. et sp. nov., MB.C.9622; bed F, *P. partitus* conodont Biozone, Eifelian, lateral view.

Figs 11–13. *Centroceras* sp., MB.C.9623; bed KMO-IV, Zlíchovian, Emsian. 11, lateral view. 12, adoral view. 13, detail of lateral view showing conch ornamentation. Scale bar represents 10 mm.

All specimens from the Filon Douze section and × 1 except where indicated by scale bar.

PLATE 4

KRÖGER, nautiloids

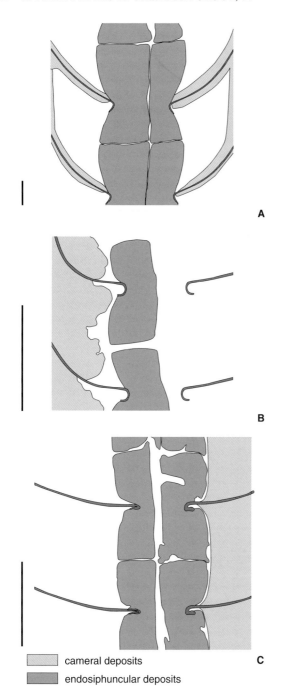

cameral deposits

endosiphuncular deposits

TEXT-FIG. 13. Camera lucida drawings of median sections of Early Devonian Ormoceratidae from Filon Douze. A, *Deiroceras hollardi* sp. nov.; MB.C.9630, *Deiroceras* Limestone, Zlíchovian. B, *Ormoceras* sp.; MB.C.9639, bed KMO-I, Pragian. C, *Metarmenoceras fatimae* sp. nov.; MB.C.9642, *Erbenoceras* Limestone, Zlíchovian. Scale bars represent 10 mm.

Remarks. Deiroceras hollardi differs from all other species of *Deiroceras* in combining a very low angle of expansion with wide chamber distance and a fine transverse ornamentation; it differs from the Silurian *D. amatum* Heritsch, 1931, in having less bent septal necks and more elongated siphuncu-

lar segments. It is possibly conspecific with *Orthoceras beyrichi* Kayser, 1878 from the Devonian of the German Harz Mountains, which is similar in general conch shape and in the shape of the siphuncle. However, the single illustration of this species (pl. 10, fig. 5) does not show the shape of septal necks and internal details of the siphuncle and the type specimen could not be located for comparison.

Genus METARMENOCERAS Flower, 1940

Type species. Metarmenoceras bilaterale Flower, 1940, from the Dalhousie beds, Early Devonian, of Gaspé, Quebec, Canada.

Diagnosis (after Flower 1940, p. 445). Slightly depressed orthocones with straight, transverse sutures; siphuncle eccentric with broadly expanded nummuloidal segments twice as wide as long; septal necks recumbent; expanded segments touch adoral and adapical surface of septa; radial canal compressed, touching inner surface of connecting ring at side of siphuncle, directed towards conch margin; radial canals bilaterally symmetric; episeptal and hyposeptal deposits.

Occurrence. Late Silurian–Early Devonian; Canada, Japan, Sardinia and Ukraine.

Metarmenoceras fatimae sp. nov.
Plate 5, figures 4–5; Text-figure 13C

Derivation of name. In honour of Fatima, daughter of Filon Douze miner Lahcan Caraoi.

Holotype. MB.C.9642 (Pl. 5, figs 4–5; Text-fig. 13C).

Type locality and horizon. Filon Douze section, *Erbenoceras* Limestone, Zlíchovian, Emsian.

Other material. 2 specimens, MB.C.9641, 9570, from the type locality and horizon; 2 specimens, MB.C.9643–44, from the *Deiroceras* Limestone at the type locality, Zlíchovian.

Diagnosis. Metarmenoceras with nearly circular cross-section and smooth conch surface; siphuncle subcentral; siphuncular segments subquadratic in median section, nearly as long as wide and strongly expanded within chambers, *c.* 0.2 of conch cross-section in diameter; diameter of septal perforation half of diameter of siphuncular segments.

Description. Three fragments have angle of expansion of 7 (holotype), 9 (MB.C.9644), and 10 (MB.C.9643) degrees. Conch very slightly curved with siphuncle on convex side of growth axis. Holotype is fragment of phragmocone 70 mm long, minimum cross-section diameter 26 mm. Second specimen (MB.C.9644) 65 mm long, minimum cross-section diameter 32 mm.

MB.C.9643 134 mm long, cross-section diameter 38–62 mm. No parts of body chamber preserved in specimens. Conch cross-section nearly circular. In holotype seven chambers occur within length of 42 mm; in second specimen ten chambers occur within length of 65 mm. Septal necks recumbent, touching adapical surfaces of septa (Pl. 5, fig. 4). Septal perforations positioned subcentrally with cross-section diameter 2.8–2.9 mm in holotype. Siphuncular segments have diameter of 5.6–6.4 mm in holotype and a subquadratic outline in median section. Connecting ring touches septal surfaces on adapical and adoral sides. Strong annular endosiphuncular deposits occur that have wide, irregularly shaped central canal. Deposits on dorsal side of siphuncle more massive and irregularly shaped. Hyposeptal and episeptal deposits occur on prosiphuncular side of conch.

Remarks. This species differs from other species of *Metarmenoceras* in having a comparatively narrow siphuncle with subquadrate segments.

Order PSEUDORTHOCERIDA Barskov, 1963

Remarks. Kröger and Isakar (2006) pointed out that Orthocerida and Pseudorthocerida are sister groups. They stated that Pseudorthocerida differ from the Orthocerida in having a conical apex with a cicatrix and cyrtochoanitic septal necks, a conclusion drawn from the observation that all cyrtochoanitic orthocones display these characters. However several morphologies exist that are intermediate between typical orthocerids and pseudorthocerids, having achoanitic or suborthochoanitic septal necks. The conical, cicatrix-bearing apex of *Suloceras melolineatum* (Niko, 1996), a spyroceratid with suborthochoanitic septal necks, shows there is a high probability that other intermediate forms have pseudorthoceratid apices. A definite determination of the higher classification of these orthocones is possible only after examination of the shape of the apex. In cases where apical characters are not known I follow the rule that taxa with typical pseudorthocerid endosiphuncular deposits (parietal deposits) and expanded siphuncular segments are placed within the Pseudorthocerida, whereas taxa with symmetrical, geisonoceroid endosiphuncular deposits are placed within the Orthocerida.

At Filon Douze, apices from Orthocerida exclusively were disovered; this is probably a sampling effect caused by the fact that Orthocerida are common and non-Orthocerida relatively rare.

Family PSEUDORTHOCERATIDAE Flower and Caster, 1935

Genus GEIDOLOCERAS gen. nov.

Type species. *Geidoloceras ouaoufilalense* sp. nov., from the Pragian of the Filon Douze section.

Derivation of name. From a contraction of *Geisonoceras* Hyatt, 1884 and *Dolorthoceras* Miller, 1931. *Geidoloceras* is superficially similar to the former but internally similar to the latter.

Other species included. None.

Diagnosis. Straight or very slightly cyrtoconic pseudorthoceridans, ornamented with rounded transverse ridges with narrow interspaces; conch cross-section nearly circular; *c.* 5 septa per distance similar to conch diameter; septal necks short cyrtocoanitic; siphuncle subcentral; siphuncular segments expanded within chambers, elongate, ovate with larger expansion at adoral third of each segment; parietal endosiphuncular deposits strongly elongated adapically.

Remarks. *Geidoloceras* differs from *Dolorthoceras* in having a transversely ornamented shell. *Geisonoceroides* Flower, 1939, differs in having concave interspaces between the transverse ridges that are similar in width to that of the ridges. Compared to *Geisonoceroides*, the ridges and interspaces of *Geidoloceras* are very closely spaced. *Fukujiceras* Niko, 1996, which is a transversely annulated pseudorthocerid with suborthochoanitic septal necks, differs from *Geidoloceras* in having endosiphuncular deposits that originate in the mid-length of segments. *Subdoloceras* gen. nov. differs in having suborthochoanitic septal necks.

Occurrence. Pragian; Morocco.

Geidoloceras ouaoufilalense sp. nov.
Plate 6, figure 6; Plate 7, figure 5; Plate 8, figure 5

Derivation of name. From the Jebel Ouaoufilal, which is close to the Filon Douze section.

Holotype. MB.C.9627 (Pl. 6, fig. 6; Pl. 7, fig. 5; Pl. 8, fig. 5).

Type locality and horizon. Filon Douze section, bed KMO-II, Pragian.

Material. 1 specimen, MB.C.9629, from bed KMO-III, Pragian, Filon Douze section.

Diagnosis. As for genus, by monotypy.

Description. Holotype with nearly circular cross-section, 21–29 mm in diameter; angle of expansion 10 degrees; ornamented with fine oblique transverse ridges (Pl. 7, fig. 5), sloping slightly towards adoral end on prosiphuncular side; ten ridges per 1 mm. In lateral view, prosiphuncular conch margin slightly convex, antisiphuncular margin slightly concave. Growth axis slightly cyrtoconic with siphuncle at convex side of curvature. Central axis of siphuncle at larger end of holotype 2 mm off growth axis. Septal distance in holotype 5 mm adapically and 8.5 mm adorally. Septal perforation at most apical septum in holotype 1.7 mm in diameter, at which point siphuncular segment expands to diameter of 3.8 mm. Siphuncular segments ovate with height to width ratio of 1.4–1.5 in holotype. Septal necks cyrtochoanitic (Pl. 8, fig. 5). Hyposeptal and episeptal deposits occur.

Genus NEOCYCLOCERAS Flower and Caster, 1935

Type species. Neocycloceras obliquum Flower and Caster, 1935, from the Lewis Run Sandstone Member, Cattaraugus Formation, Late Devonian of Lewis Run, Pennsylvania, USA.

Diagnosis (after Sweet 1964*b*, p. K244). Slender, uncompressed to very slightly depressed orthocones with large sinuous, slightly oblique superficial undulations; sutures more oblique than undulations, projected adaperturally on prosiphuncular side; sutural obliquity increasing with age; sutures with broad, low dorsal saddle, high, conspicuous ventral saddle, and lateral lobes; siphuncle between centre and venter; segments nummuloidal, occupied by continuous lamellar endosiphuncular deposit, thickest at widest part of siphuncle and thin in region of neck.

Remarks. This genus was regarded as questionable by Sweet (1964*b*), probably because of the poor knowledge of the internal characters of the type. Flower and Caster (1935) provided no description of the septal neck shape of *N. obliquum*. However, it is known that the siphuncle is clearly inflated and it can be deduced from the expanded siphuncular segments that its septal necks are suborthochoanitic or cyrtochoanitic. This is probably the reason why Flower (1939), Shimansky (1962) and Sweet (1964*b*) placed the genus within the Pseudorthoceratidae. Zhuravleva (1978, p. 150) stated in the generic diagnosis of *Neocycloceras* that the septal necks are cyrtochoanitic, but provided no evidence for this. Herein, I follow the established practice and assign to *Neocycloceras* species with an ornament similar to that of *Monomuchites* Wilson, 1961, but with obliquely lobate sutures, broadly inflated siphuncular segments and septal necks of pseudortheratid aspect. *Neocycloceras* differs from *Reticycloceras* Gordon, 1960 in lacking a longitudinal ornamentation.

Occurrence. Early Devonian–Late Permian; world-wide.

Neocycloceras? termierorum sp. nov.
Plate 5, figure 6; Plate 6, figure 4

Derivation of name. In honour of Geneviève and Henri Termier for their lifelong engagement in the palaeontology of Morocco.

Holotype. MB.C.9677 (Pl. 5, fig. 6; Pl. 6, fig. 4).

Type locality and horizon. Filon Douze section, bed KMO-IV, Zlíchovian, Emsian.

Other material. 2 specimens, MB.C.9656, 9673, from bed F, early *O. costatus* conodont Biozone, middle Eifelian, Filon Douze section.

Diagnosis. Neocycloceras with angle of expansion of *c.* 8 degrees and comparatively narrowly undulating shell; more than four undulations per distance similar to cross-section diameter; undulations slightly oblique, sloping in adoral direction on prosiphuncular side forming shallow lobe; sutures more oblique than undulations, projected adaperturally on prosiphuncular side; one undulation between two septa; septal spacing *c.* 0.16 of cross-section diameter; sutures slightly oblique, projecting adaperturally on prosiphuncular side, forming conspicuous ventral saddle and shallow lateral saddles; siphuncle eccentric with septal perforation 0.25 of conch cross-section; septal necks cyrtochoanitic.

Description. Holotype is fragment of phragmocone, 15 mm long, nearly circular cross-section with diameter 17.8–19.8 mm and angle of expansion 7.6 degrees. Shell surface ornamented with slightly obliquely transverse undulations. Spacing of undulations 4 mm, which is similar to that of septa. Undulations have sharp, narrow ridges and wide, rounded valleys; depth between ridges and valleys 0.25 mm. Undulations slope slightly in adoral direction on prosiphuncular side. Ridges positioned on adoral surface of septum on antisiphuncular side and on adapical side of succeeding septum on prosiphuncular side (Pl. 5, fig. 6). Septa slightly oblique, forming shallow lateral lobes and distinctive saddle on prosiphuncular side. Septal perforation at most adoral septum 2 mm wide and positioned with centre 8.9 mm from conch margin. Septal necks suborthochoanitic, length *c.* 0.7 mm (0.35 of diameter of septal perforation). Traces of connecting rings in some chambers reveal strongly expanded siphuncular segments.

Remarks. N? termierorum differs from other species of *Neocycloceras* in having a comparatively narrow undulation and a comparatively central siphuncle; it differs from *N. obliquum* in its coincidence of chamber height and undulation. In *N. obliquum c.* 3 chambers occur in the span of one annulation. The septal necks of *N? termierorum* are suborthochoanitic, but clearly longer than in Dawsonoceratidae. The new species is questionably assigned to *Neocycloceras* because at present the type of this genus is inadequately known.

Genus PROBATOCERAS Zhuravleva, 1978

Type species. Probatoceras crassum Zhuravleva, 1978, from the Frasnian of Sargajevo, South Urals, Russia.

Emended diagnosis. Short orthocones or slight cyrto-cones with circular or slightly compressed cross-section; shell with distinct transverse lirae; septa closely or medium spaced; sutures slightly oblique, sloping towards aperture on prosiphuncular side; siphuncle central or subcentral with expanded segments; septal necks suborthochoanitic; endosiphuncular and cameral deposits not known.

Remarks. In the original diagnosis (Zhuravleva 1978, p. 140) the septal necks were described as 'rather long cyrtochoanitic' and the siphuncle as 'central'. However, from the figures of the type it is evident that the septal necks are suborthochoanitic in the strict sense (Kröger and Isakar 2006). Moreover, the illustration of the type species shows that the siphuncle is subcentral.

Occurrence. Pragian?–Frasnian; Czech Republic, Morocco and Ural Mountains (Russia).

Probatoceras? sp
Plate 6, figure 12; Plate 8, figure 4

Material. 1 specimen, MB.C.9651, from KMO-II, Pragian, Filon Douze section.

Description. Fragment of phragmocone 24 mm long, cross-section circular, diameter 7.6–10.3 mm (angle of expansion 6.4 degrees). Conch surface not preserved. Sutures slightly oblique, slope in adoral direction on prosiphuncular side. Fragment has eight complete chambers. Distance between two successive septa *c.* 0.2 of conch cross-section; septal perforation subcentral, diameter *c.* 0.1 of conch cross-section. Septal necks suborthochoanitic, length 0.3 of septal perforation diameter.

Remarks. This specimen cannot be assigned to *Probatoceras* with certainty because the conch surface, which would yield diagnostic characters, is not preserved. However, the close septal spacing, the wide angle of expansion, and the shape of the septal necks of the specimen are diagnostic characters of *Probatoceras*.

Genus SUBDOLOCERAS gen. nov.

Type species. *Subdoloceras tafilaltense* sp. nov., from the Daleijan of the Filon Douze section.

Derivation of name. Latin, *sub*, beneath, and resemblance to *Dolorthoceras* Miller, 1931, to indicate the presence of suborthochoanitic septal necks.

Other species included. *Subdoloceras atrouzense* sp. nov.; *S. engeseri* sp. nov.

Diagnosis. Straight pseudorthocerids with transversely lirate shell; conch cross-section nearly circular; 3–4 septa at distance similar to conch diameter; septal necks suborthochoanitic; siphuncle subcentral; siphuncular segments expanded within chambers, elongate-ovate with larger expansion at adoral third of each segment; parietal endo-siphuncular deposits.

Remarks. The presence of suborthochoanitic septal necks, siphuncular segments of the *Dolorthoceras* type (see Flower 1939, fig. 3) and a narrow transverse ornamentation are the diagnostic characters of *Subdoloceras*. It differs from *Dolorthoceras* in having a transverse ornamentation and suborthochoanitic septal necks, and from *Geidoloceras* in having suborthochoanitic septal necks. In *Dolorthoceras* the septal necks of early growth stages are orthochoanitic, becoming cyrtochoanitic in later growth stages. The type species of *Subdoloceras* is known from early ontogenetic parts of the phragmocone, which also display suborthochoanitic septal necks. *Geidoloceras* has cyrtochoanitic septal necks.

Occurrence. Pragian–Daleijan; Morocco.

Subdoloceras tafilaltense sp. nov.
Plate 5, figure 9; Plate 6, figure 3; Plate 8, figure 3

Derivation of name. From Tafilalt, in which the type locality of the species is located.

Holotype. MB.C.9695 (Pl. 5, fig. 9; Pl. 6, fig. 3).

Type locality and horizon. Filon Douze section, bed KMO-IV, Zlichovian, Emsian.

Other material. Known only from the holotype.

Diagnosis. *Subdoloceras* with straight, constantly widening conch, angle of expansion of 12 degrees; cross-section circular; shell surface ornamented with fine, slightly obliquely transverse lirae; septal spacing *c.* 0.35 of cross-section diameter; septa directly transverse or slightly oblique; siphuncle subcentral with diameter of septal perforation 0.08 of conch cross-section; siphuncular segments expanded with point of largest concavity at adoral third of segment; ratio of diameter to length of siphuncular segments *c.* 0.5; septal necks suborthochoanitic.

Description. Holotype is straight fragment of phragmocone, 40 mm long, cross-section diameter 10.8–19 mm. Shell surface ornamented with fine, slightly obliquely transverse lirae that slope in adoral direction on prosiphuncular side of conch, four per 1 mm on adoral part of fragment. Septal spacing between 6.2 mm at adoral end and 3.9 mm at apical end. Siphuncle subcentral with diameter of septal perforation of 1 mm where cross-section of fragment is 12 mm. Septal necks suborthochoanitic throughout entire length. Eight septa preserved in holotype. Siphuncular segments expanded with point of largest concavity at adoral third of segment (Pl. 8, fig. 3). Maximum siphuncular diameter at most adapical segment 1.5 mm. No traces of endosiphuncular or cameral deposits occur.

Remarks. This species differs from *S. engeseri* sp. nov. in having a lower expansion rate and a less eccentric siphuncle.

Subdoloceras atrouzense sp. nov.
Plate 6, figure 1; Plate 7, figure 1; Plate 8, figure 1

Derivation of name. From the El Atrouz mine to the east of the Filon Douze section.

Holotype. MB.C.9595 (Pl. 6, fig. 1; Pl. 7, fig. 1; Pl. 8, fig. 1).

Type locality and horizon. Filon Douze section, bed KMO-II, Pragian.

Other material. 3 specimens, MB.C.9681–82, 9645, from beds KMO-II and KMO-III, Pragian, Filon Douze section.

Diagnosis. *Subdoloceras* with slightly cyrtoconic conch; angle of expansion 15 degrees; cross-section circular or slightly compressed; ornamented with fine, obliquely transverse rounded ridges; septal spacing *c.* 0.2 of cross-section diameter; septa slightly oblique, sloping in adoral direction on prosiphuncular side; siphuncle subcentral with diameter of septal perforation 0.1 of conch cross-section; septal necks suborthochoanitic.

Description. Holotype is fragment of phragmocone 23 mm long, cross-section diameter 8–14.5 mm, ornamented with slightly obliquely transverse ridges that slope towards adoral end on prosiphuncular side (Pl. 7, fig. 1), ten ridges per 1 mm. In lateral view conch margins slightly concave. Growth axis very slightly curved with subcentral siphuncle on convex side of growth axis. Septal spacing 2.9 mm at adoral end and 1.5 mm at apical end. Septal perforation of 1.5 mm at most adoral septum. Septal necks suborthochoanitic (Pl. 7, fig. 1). Holotype shows no traces of connecting ring or endosiphuncular or cameral deposits.

MB.C.9681 is fragment of phragmocone 26 mm long; cross-section nearly circular, diameter *c.* 12–18 mm (angle of expansion 15 degrees). Conch margin in median section slightly convex. Surface transversely ornamented with ten ridges per 1 mm.

Two additional specimens have poorly preserved shell surfaces. MB.C.9645 19 mm long, diameter 19–22 mm (angle of

expansion 9 degrees). Distance between septa 4 mm (septal distance 0.18 of cross-section diameter). Septal perforation slightly subcentral. Septal necks suborthochoanitic. MB.C.9682 12 mm long, cross-section diameter 4.5–7 mm (angle of expansion 12 degrees). Distance between septa 1.6 mm (septal distance 0.22 of cross-section diameter). Septal perforation subcentral. Septal necks cyrtochoanitic. Siphuncular segments strongly expanded within chambers, ovate and adnate on adoral surfaces of septa; length to height ratio 1.4.

Remarks. This species of *Subdoloceras* is characterized by its high expansion rate.

Subdoloceras engeseri sp. nov.
Plate 6, figure 5; Plate 7, figure 2; Plate 8, figure 7

Derivation of name. In honour of Theo Engeser, Berlin, for his compilation of a comprehensive taxonomic database of non-ammonoid cephalopods.

Holotype. MB.C.9654 (Pl. 8, fig. 7).

Type locality and horizon. Filon Douze section; bed KMO-I, Pragian.

Other material. 15 specimens, MB.C.9657–72, from bed KMO-II, Pragian, Filon Douze section.

Diagnosis. *Subdoloceras* with angle of expansion of 7–9 degrees; cross-section circular or slightly depressed; ornamented with fine, obliquely transverse, rounded ridges with narrow interspaces; ridges slope in adoral direction on prosiphuncular side; septal spacing 0.3–0.4 of cross-section diameter; septa slightly oblique, sloping in adoral direction on prosiphuncular side; obliquity of ornamentation stronger than that of septa; sutures straight; siphuncle subcentral with septal perforation 0.1 of conch cross-section; siphuncular segments expanded with maximum diameter *c.* 0.14 of conch cross-section; septal necks suborthochoanitic; parietal endosiphuncular deposits strongly elongated in direction of apex, slightly touching connecting ring at septal necks; hyposeptal and episeptal deposits.

Description. Holotype with nearly straight growth axis; nearly circular cross-section 15–19 mm in diameter; angle of expansion of conch 9 degrees, ornamented with fine, transverse oblique lirae, sloping slightly towards adoral end on presiphuncular side (Pl. 7, fig. 2). Twenty lirae per 1 mm. Angle of expansion 7 degrees in MB.C.9658, with maximum cross-section diameter of 15 mm, 8 degrees in MB.C.9657, with maximum cross-section diameter of 16 mm. In holotype, septal distance between 5 mm adapically and 6 mm adorally, central axis of siphuncle apically at distance of 1.4 mm from growth axis, diameter of septal perforation at most adoral septum 1.6 mm. Siphuncular segments expanded with length to height ratio of 1.4 in MB.C.9659, 1.5 in MB.C.9660. Septal necks suborthochoanitic (Pl. 8, fig. 7). In smaller fragments, siphuncular tube appears less expanded with ratio of height to length of siphuncular segments 2.0. Parietal endosiphuncular deposits occur in MB.C.9658 at a cross-section diameter <11 mm.

Remarks. This species differs from *S. tafilaltense* sp. nov. in having a relatively low expansion rate and a more eccentric siphuncle.

Genus SUBORMOCERAS gen. nov.

Type species. *Subormoceras erfoudense* sp. nov., from the Zlíchovian (Emsian) of the Filon Douze section.

Derivation of name. Latin, *sub*, under, referring to the suborthochoanitic septal necks of this genus, which otherwise is similar to *Ormoceras*.

Other species included. *Ormoceras dobrovljanense* Kiselev, *in* Balashov and Kiselev 1968; *O. rashkovense* Kiselev, *in* Balashov and Kiselev 1968; *O. skalaense* Kiselev, *in* Balashov and Kiselev 1968 (all species combined with *Ormoceras* below: see 'List of new combinations'); *S. rissaniense* sp. nov.

Diagnosis. Straight conch with nearly circular cross-section; angle of expansion <10 degrees; sutures straight, transverse; septal distance small, <0.3 of conch cross-section; siphuncle central–eccentric; siphuncular segments expanded within chambers, elliptical or subcircular in median section, without area of adnation; septal perforation *c.* 0.1–0.15 of conch cross-section; septal necks suborthochoanitic; endosiphuncular deposits not known; episeptal deposits.

Remarks. This genus differs from *Ormoceras* in having suborthochoanitic septal necks and there are either no endosiphuncular deposits or they are restricted to the very adapical parts of the phragmocone of mature specimens. *Subdoloceras* differs in having a narrower siphuncle and a wider septal spacing.

Occurrence. Ludfordian to Eifelian; Morocco and Ukraine.

Subormoceras erfoudense sp. nov.
Plate 8, figure 6

Derivation of name. From Erfoud in south-east Tafilalt.

Holotype. MB.C.9683 (Pl. 8, fig. 6).

Type locality and horizon. Filon Douze section, *Erbenoceras* Limestone, Zlíchovian, Emsian.

Other material. 1 specimen, MB.C.8684, from the type locality and horizon; 3 specimens, MB.C.9571.1–3, from bed F, *P. partitus* conodont Biozone, Eifelian, the type locality.

Diagnosis. Slender, straight conch with nearly circular cross-section; angle of expansion very low; sutures straight, transverse; septal distance 0.3 of conch cross-section; siphuncle nearly central; siphuncular segments subglobular with width to height ratio of *c.* 1.0; siphuncular diameter *c.* 0.12 of cross-section; septal necks suborthochoanitic, forming rounded brim; length of septal necks *c.* 0.15 of septal distance, 0.3 of diameter of septal perforation; episeptal deposits.

Description. Holotype 50 mm long, with slightly compressed cross-section 36–38 mm in diameter (angle of expansion *c.* 1 degree). Surface poorly preserved but apparently smooth. Fragment displays seven complete chambers. Distance between two most adoral septa 13 mm, and between most adapical 8 mm. Septal curvature with depth of 12 mm at most adoral septum. Sutures straight, transverse. Septal perforation 5 mm in cross-section. Septal necks suborthochoanitic, form *c.* 1.5-mm-long, rounded brims. Siphuncle subcentral. Centre of septal perforation 21 mm from conch margin. Siphuncular segments expanded within chambers, height and width about equal. Endosiphuncular and cameral deposits unknown.

MB.C.8684 33 mm long, maximum conch cross-section 35 mm, with three complete chambers preserved 10, 8.9 and 8.0 mm in length from adoral to adapical end. Septal perforation nearly central, 4 mm in diameter. Episeptal deposits present.

Remarks. This species is very similar to *Ormoceras dobrovljanense* Kiselev, *in* Balashov and Kiselev 1968; from which it differs only in having a more slender siphuncle that is positioned closer to the growth axis.

Subormoceras rissaniense sp. nov.
Plate 4, figure 10; Plate 8, figure 9

Derivation of name. From Rissani, in eastern Tafilalt.

Holotype. MB.C.9678 (Pl. 8, fig. 9).

Type locality and horizon. Filon Douze section, bed PI, early *O. costatus* conodont Biozone, middle Eifelian.

Other material. 1 specimen, MB.C.9622, from bed F, *P. partitus* conodont Biozone, Eifelian, Filon Douze section.

Diagnosis. Straight conch with nearly circular cross-section; angle of expansion below 9 degrees; sutures straight, transverse; septal distance 0.15 of conch cross-section; siphuncle subcentral; siphuncular segments expanded within chambers; siphuncular diameter *c.* 0.1 of cross-section; septal necks suborthochoanitic, form angular brims; length of septal neck *c.* 0.3 of septal distance, 0.5 of diameter of septal perforation; endosiphuncular and cameral deposits unknown.

Description. Holotype 50 mm long, slightly compressed cross-section, diameter of 42–43 mm (angle of expansion 9 degrees). Conch straight, margins slightly convex in median section. Surface poorly preserved in holotype but second specimen has a smooth surface with faint, irregularly spaced, growth lines. Holotype displays nine complete chambers. Distance between two most adoral septa 5.4 mm, between most adapical chambers 5.5 mm. Septal curvature with depth of 3.8 mm at most adoral septum. Sutures straight, transverse. Septal perforation of most adoral septum 3 mm in cross-section. Septal necks suborthochoanitic, form *c.* 1.5-mm-long, angular brims. Siphuncle subcentral. Centre of septal perforation 15.6 mm from conch margin at most adoral septum. Siphuncular segments expanded within chambers but their shape is not known. Endosiphuncular and cameral deposits unknown.

Second specimen ornamented with very fine, directly transverse, growth lines 63 mm long, conch cross-section 33–40 mm in diameter. Conch margins slightly convex in median section. Only most adoral septum completely preserved, showing septal perforation 4.4 mm in cross-section and suborthochoanitic septal necks, which form angular brims.

Remarks. This species is very distinctive because of the convex shape of the conch margins in median section and the angular shape of the suborthochoanitic septal necks.

Subormoceras sp.
Plate 4, figures 6–7; Plate 8, figure 10

Material. One specimen, MB.C.9674 from bed OSP-I, Přídolí, Filon Douze section.

Description. Conch fragment 40 mm long; cross-section 31–35 mm in diameter (angle of expansion 6 degrees); margins slightly convex in median section; cross-section slightly compressed, ratio of conch width to height 0.93, subtriangular with flattened antisiphuncular side and rounded lateral side. Conch surface poorly preserved but smooth or only slightly ornamented, without undulations. Sutures form shallow lateral saddles and wide lobe on prosiphuncular side. Septal distance between 5.3 and 4.1 mm (0.16–0.13 of cross-section diameter). Septa imploded and fragmentary, providing no information on exact position of siphuncle, which is not marginal. Diameter of septal perforation 4 mm (0.14 of conch cross-section). Septal necks suborthochoanitic to orthochoanitic, 1.4 mm long (Pl. 8, fig. 10).

Remarks. The very close septal spacing and the suborthochoanitic septal necks are diagnostic characters of *Subor-*

EXPLANATION OF PLATE 5

Median sections of Oncocerida, Actinoceratida and Pseudorthocerida from the Filon Douze section.

Fig. 1. *Mutoblakeoceras inconstans* gen. et sp. nov., MB.C.9584; bed PI, early *O. costatus* conodont Biozone, middle Eifelian; detail of siphuncular structure, conch margin on the right; note the concave appearance of siphuncular segments, orthochoanitic septal necks and endosiphuncular lamellae.

Fig. 2. Indet. Nothoceratidae sp. A., MB.C.9635; bed PI, early *O. costatus* conodont Biozone, middle Eifelian; detail of siphuncular structure, conch margin towards the left; note the concave appearance of siphuncular segments, orthochoanitic septal necks and endosiphuncular lamellae.

Fig. 3. *Ormoceras* sp., MB.C.9639; bed KMO-I, Pragian; note the cyrtochoanitic septal necks and the irregularly shaped, massive ventral endosiphuncular annuli.

Figs 4–5. *Metarmenoceras fatimae* sp. nov., MB.C.9674, holotype; *Erbenoceras* Limestone, Zlíchovian, Emsian. 4, detail of siphuncular structure; note the recumbent septal necks, the large area of adnation of the connecting ring, and the massive endosiphuncular deposits. 5, entire specimen.

Fig. 6. *Neocycloceras*? *termierorum* sp. nov., MB.C.9677, holotype; bed KMO-IV, Daleijan, Emsian; detail of septal perforation showing cyrtochoanitic septal necks.

Figs 7–8. *Deiroceras hollardi* sp. nov., MB.C.9630; *Deiroceras* Limestone, Zlíchovian, Emsian. 7, entire specimen; note the large, elliptical siphuncular segments. 8, detail of siphuncular structure showing suborthochoanitic septal necks and massive endosiphuncular annuli.

Fig. 9. *Subdoloceras tafilaltense* gen. et sp. nov., MB.C.9695, holotype; bed KMO-IV, Daleijan, Emsian; detail of septal perforation showing suborthochoanitic septal necks.

Fig. 10. *Diagoceras* sp., MB.C.9579, bed PI, early *O. costatus* conodont Biozone, middle Eifelian; section shows oblique, narrowly spaced septa and suborthochoanitic septal necks.

Scale bars represent 1 mm; figures without a scale bar are × 1.

PLATE 5

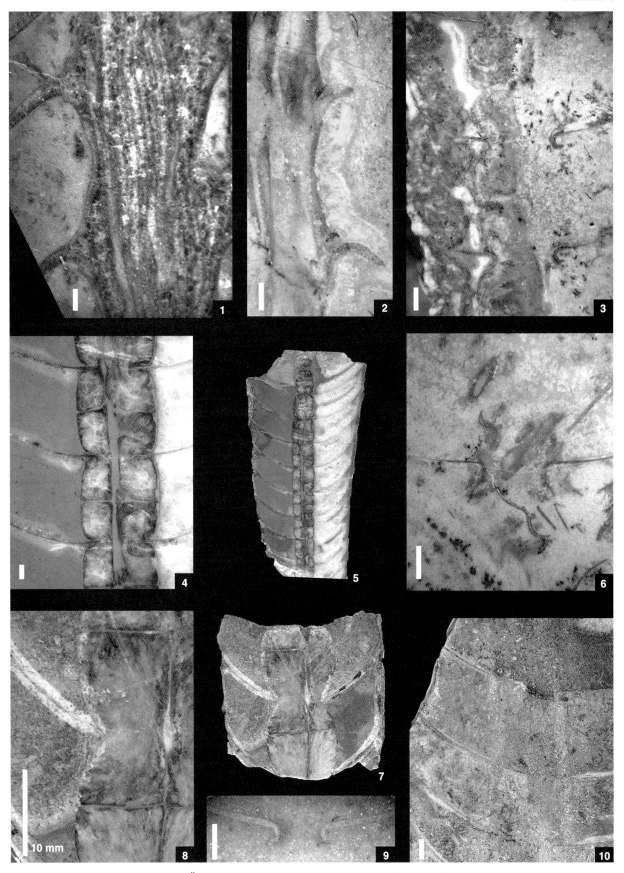

moceras. However, the exact position of the septal perforation and the shape of the siphuncular segments are not known, making a specific determination of the fragment impossible.

Family SPYROCERATIDAE Shimizu and Obata, 1935*a*

Remarks. I follow the diagnosis of Sweet (1964*b*, p. K246) in assigning to the Spyroceratidae orthocones that have siphuncular segments and endosiphuncular deposits with the aspect of *Dolorthoceras*. The septal necks of *Spyroceras* are cyrtochoanitic; those of *Dolorthoceras* are cyrtochoanitic in adult growth stages and suborthochoanitic in juvenile stages (see Flower 1939, fig. 3). The septal necks and siphuncular segments of *Cancellspyroceras* gen. nov. and *Suloceras* Manda, 2001, resemble those of juvenile *Dolorthoceras*. I place these two genera within the Spyroceratidae because, in addition to their *Dolorthoceras*-like septal neck shape, they have *Spyroceras*-like ornamentation.

Genus SPYROCERAS Hyatt, 1884

Type species. *Orthoceras crotalum* Hall, 1879, from the Skaneateles Shale, Hamilton stage, Devonian, of Pratt's Fall, Onondaga County, New York, USA.

Diagnosis (after Sweet 1964*b*, p. K246). Undulating orthocones with straight, transverse sutures, transverse or slightly oblique undulations and faintly cyrtoconic apices; surface ornamented with longitudinal lirae conspicuous from earliest stage and with transverse elements; siphuncle central or slightly ventrally shifted from centre; siphuncular segments and endosiphuncular deposits of *Dolorthoceras* type (Flower 1938); septal necks cyrtochoanitic; cameral deposits developing later than in other Pseudorthoceratidae, hence confined to more apical regions of conch.

Occurrence. The time range and geographical occurrence of the genus is difficult to assess because it serves as a 'wastebasket' taxon for orthocones with similar ornamentation but different internal characters. It seems to be restricted to the Devonian and has a world-wide distribution.

Spyroceras cyrtopatronus sp. nov.
Plate 6, figure 9; Plate 7, figure 4; Plate 8, figures 8, 11

Derivation of name. Greek, *kyrtos*, curved, with reference to the bent growth axis of this species, which is otherwise similar to *S. patronus* (Barrande, 1866).

Holotype. MB.C.9676 (Pl. 8, figs 8, 11).

Type locality and horizon. Filon Douze section, bed KMO-III, Pragian.

Other material. 1 specimen, MB.C.9649, from the type locality and horizon.

Diagnosis. Cyrtoconic *Spyroceras* with angle of expansion of 12–15 degrees; cross-section slightly compressed; conch with straight undulations; surface ornamented with *c.* 150–200 distinctive, raised, irregularly spaced, longitudinal lirae around circumference and numerous faint, directly transverse lirae; approximately two transverse lirae at distance similar to distance between two longitudinal lirae; septa directly transverse at distance of *c.* 0.25 of cross-section diameter; septal perforation *c.* 0.1 of conch cross-section diameter; septal necks cyrtochoanitic; siphuncle subcentral on convex side of conch curvature.

Description. Holotype is fragment of phragmocone 32 mm long; compressed cross-section with width to height ratio of 0.92, diameter 6–13 mm (angle of expansion 12 degrees). Conch cyrtoconic with siphuncle slightly shifted from centre towards convex side of conch curvature. MB.C.9649 39 mm long, cross-section diameter 6.5–17 mm (angle of expansion 15 degrees). Surface of holotype poorly preserved with distance between longitudinal lirae at adoral end averaging 0.3 mm, transverse lirae at distances of 0.1 mm; distance between longitudinal lirae variable. MB.C.9649 has, at adoral end (cross-section diameter 17 mm), irregularly spaced longitudinal lirae at distances of *c.* 0.3 mm (Pl. 6, fig. 9; Pl. 7, fig. 4) and directly transverse shallow undulations at distances of *c.* 4 mm. Holotype has 12 septa with chamber height of 3 mm at adoral end, 1.5 mm at adapical end. Shallow septal concavity with depth of curvature 2.9 mm at most adoral chamber. Septal perforation at most adoral septum of holotype 1 mm wide. Septal necks cyrtochoanitic (Pl. 8, figs 8, 11). Siphuncular segments slightly expanded within chambers; siphuncle subcentral. Siphuncular segments with height to width ratio of *c.* 1.7.

Remarks. This species differs from *S. patronus* in having a higher expansion rate, a cyrtoconic conch and more closely spaced longitudinal ornamentation, and from *S. latepatronus* sp. nov. in having a lesser angle of expansion and less regular longitudinal ornamentation.

Spyroceras fukuijense? Niko, 1996
Plate 6, figure 11; Plate 7, figure 9; Plate 9, figure 4

*1996 *Spyroceras fukuijense* Niko, p. 350, figs 3, 3–11, 4, 1–5.

Material. 2 specimens, MB.C.9776–77, from KMO-III, Pragian, Filon Douze section.

Diagnosis (after Niko 1996, p. 350). Slightly cyrtoconic *Spyroceras* with low angle of expansion; cross-section circular or slightly depressed; conch with straight undulations; undulations less distinctive in later growth stages; ornamented with 25–33 distinctive, raised, longitudinal lirae around circumference and numerous faint, directly transverse lirae; additionally, reticulate ornamentation occurs on body chamber; siphuncle slightly subcentral on concave side of conch curvature.

Description. Larger specimen (MB.C.9777) 17 mm long, cross-section diameter 16–18 mm (angle of expansion 7 degrees). Surface ornamented with 26 prominent longitudinal lirae (Pl. 6, fig. 11). Undulations directly transverse with distances of *c.* 5 mm. Distance between transverse striae 3–4 mm. Septal distance 4.5 mm. Septal perforation slightly shifted from growth axis towards concave side of conch curvature with diameter of 2.2 mm at cross-section diameter of 17 mm (MB.C.9777). Septal necks suborthochoanitic (Pl. 9, fig. 4).

Remarks. The specimens described above display the diagnostic characters of the phragmocone of *S. fukuijense*. However, fragments of the adult body chamber of this species are not known from Filon Douze. Because the body chamber of *S. fukuijense* shows a conspicuous reticulate ornamentation that has diagnostic value, the Filon Douze fragments cannot be assigned to the species with certainty.

Occurrence. Lochkovian to Pragian; Japan and Morocco.

Spyroceras latepatronus sp. nov.
Plate 6, figure 7; Plate 7, figure 3; Plate 9, figures 1–2

Derivation of name. Latin, *latus*, wide, referring to the wide angle of expansion of this species compared with *S. patronus*.

Holotype. MB.C.9653 (Pl. 6, fig. 7; Pl. 7, fig. 3).

Type locality and horizon. Filon Douze section, bed KMO-I, Pragian.

Other material. 2 specimens, MB.C.9779, 9694, from the type locality and horizon.

Diagnosis. Slightly cyrtoconic *Spyroceras* with variable angle of expansion of more than 15 degrees; cross-section circular or slightly depressed; conch with oblique undulations, sloping in adoral direction on convex side of conch curvature; surface ornamented with *c.* 60 distinctive, raised, longitudinal lirae around circumference and numerous faint, directly transverse lirae; approximately three transverse lirae at distance similar to distance between two longitudinal lirae; septa directly transverse at

distance of *c.* 0.2 of cross-section diameter; septal perforation width *c.* 0.15 of cross-section diameter; septal necks cyrtochoanitic, siphuncle subcentral on convex side of conch curvature.

Description. Holotype 22 mm long, cross-section diameter 8–13 mm (angle of expansion 16 degrees). Cross-section nearly circular. Conch slightly cyrtoconic with siphuncle slightly shifted from centre towards convex side of conch curvature, slightly obliquely undulating (Pl. 6, fig. 7). Distance between undulations 3.5 mm. Longitudinal lirae 0.3 mm apart, transverse lirae 0.1 mm apart. Vertical lirae as distinctive as transverse lirae. In longitudinal section on prosiphuncular side slightly concave, on antisiphuncular side slightly convex. Distance between chambers 2.5–3.5 mm. Septal perforation has diameter of 0.6 mm with cyrtochoanitic septal necks and expanded siphuncular segments. Siphuncular segments with height to width ratio of *c.* 1.7. Episeptal and mural deposits present.

MB.C.9779 35 mm long, maximum diameter 21 mm. Conch fragmented; angle of expansion >13 degrees. Conch surface poorly preserved but about three transverse lirae at distance similar to distance between longitudinal lirae. Septal distance at adoral end 4 mm. Septal perforation 1.5 mm in diameter at apical end. Septal necks cyrtochoanitic (Pl. 9, fig. 1).

Remarks. This species differs from *S. patronus* in having a higher expansion rate and more widely spaced transverse ornamentation.

Spyroceras patronus (Barrande, 1866)
Plate 7, figure 6; Plate 9, figure 3

*1866–70 *Orthoceras patronus* Barrande, pl. 228, figs 5–6; pl. 262, figs 12–13; pl. 275, figs 20–28; pl. 445, figs 9–11.
   1885 *Orthoceras patronus* Barrande; Tshernyshev, p. 10, pl. 1, figs 10–11.
   1978 *Spyroceras patronus* (Barrande); Zhuravleva, pp. 89, 94.
   1984 *Orthoceras patronus* Barrande; Dzik, p. 125, text-fig. 49.17.

Material. 1 specimen, MB.C.9652, from bed KMO-I and 1 specimen, MB.C.9778, from KMO-II, Pragian, Filon Douze section.

Diagnosis. Slightly cyrtoconic *Spyroceras* with variable angle of expansion of 6–12 degrees; cross-section circular or slightly depressed; adult body chamber with cross-section diameter of *c.* 35 mm, adorally slightly contracted; conch with oblique undulations, sloping in adoral direction on convex side of conch curvature; undulations less distinctive in later growth stages and disappear at adult body chamber; surface ornamented with *c.* 60 distinctive, raised, longitudinal lirae around circumference and numerous faint, directly transverse lirae; *c.* 5–10 lirae at distance similar to distance between two longitudinal lirae; septa directly transverse at distance of *c.* 0.2 of conch cross-section; septal perforation width *c.* 0.15 of conch cross-section; septal necks cyrtochoanitic; siphuncle subcentral on convex side of conch curvature; siphuncular segments expanded within chambers.

Description. Specimen MB.C.9778 32 mm long, cross-section diameter 20–24 mm (angle of expansion 10 degrees). MB.C.9652 36 mm long, cross-section diameter 17–22 mm (angle of expansion 8 degrees). Distance between longitudinal striae 1 mm, between transverse striae 0.2 mm, in both specimens. Septal perforation subcentral on convex side of conch curvature with diameter of 2.5 mm at cross-section diameter of 20 mm (MB.C.9697). Septal necks short, suborthochoanitic to cyrtochoanitic (Pl. 9, fig. 3).

Remarks. The specific diagnosis refers to the excellent figures of 'Orthoceras' patronus of Barrande (1868), which include details of the siphuncle, septal necks and adult characters. *Spyroceras patronus* differs from *Suloceras pulchrum* (Barrande, 1868) in having a slightly cyrtoconic conch and a higher expansion rate. The ornamentation of *S. pulchrum* also differs in consisting of alternating prominent and subordinate longitudinal lirae. The fragments from Filon Douze here referred to *S. patronus* differ from Barrande's specimens in having a larger angle of expansion. These variations are regarded as intraspecific. However, two specimens from the Přídolí of Filon Douze that are otherwise similar to *S. patronus* conspicuously deviate from the moderate expansion rate of the type figured in Barrande (1868, pl. 275, figs 20, 24–26); these are assigned to *S. latepatronus* sp. nov. *Spyroceras patronus* differs from *S. fukuijense* Niko, 1996 in having a higher number and less prominent longitudinal striae and a wider angle of expansion.

Occurrence. Pragian; Czech Republic, Kazakhstan and Morocco.

Spyroceras? sp. A
Plate 7, figure 7

Material. 1 specimen, MB.C.9698, from bed KMO-III, Pragian, Filon Douze section.

Description. Fragment slightly cyrtoconic, 25 mm long, cross-section slightly compressed 20–24 mm in diameter (angle of expansion 9 degrees). Conch shallowly undulating with distance between undulations 5 mm, longitudinal raised lirae 1 mm apart (75 lirae occur across circumference), directly transverse, regularly spaced lirae 0.3 mm apart. Undulations describe shallow lateral lobes. Distance between septa 3.5 mm. Septal concavity shallow. Septal perforation subcentral with diameter 1.1 mm at adapical end of fragment (0.16 of conch-cross-section).

Remarks. This fragment displays the external characters of *Spyroceras* but the internal characters are unknown. The species differs from *S. patronus* in having 75, as opposed to *c.* 60, longitudinal lirae around the circumference.

Spyroceras? sp. B
Plate 7, figure 8

Material. 1 specimen, MB.C.9699, from bed KMO-III, Pragian, Filon Douze section.

Description. Fragment slightly cyrtoconic, 30 mm long, slightly compressed cross-section with diameter of 13–16 mm (angle of expansion 6 degrees). Conch strongly undulating with distance between undulations 4 mm; irregularly spaced, longitudinal raised lirae 0.4 mm apart (*c.* 100 lirae occur around circumference); directly transverse, regularly spaced lirae 0.1 mm apart. Septal concavity shallow.

Remarks. The fragment displays the external characters of *Spyroceras* but the internal characters are unknown.

EXPLANATION OF PLATE 6

Fig. 1. *Subdoloceras atrouzense* gen. et sp. nov., MB.C.9595, holotype; bed KMO-II, Pragian; lateral view with epizoan.

Fig. 2. *Sphooceras truncatum* (Barrande, 1860), MB.C.9801; bed OP-M, Ludfordian, Ludlow; lateral view.

Fig. 3. *Subdoloceras tafilaltense* gen. et sp. nov., MB.C.9695, holotype; bed KMO-IV, Daleijan, Emsian; lateral view.

Fig. 4. *Neocycloceras*? *termierorum* sp. nov., MB.C.9677, holotype; bed KMO-IV, Daleijan, Emsian; lateral view.

Fig. 5. *Subdoloceras engeseri* gen. et sp. nov., MB.C.9657; bed KMO-I, Pragian; lateral view.

Fig. 6. *Geidoloceras ouaoufilalense* gen. et sp. nov., MB.C.9627, holotype; bed KMO-III, Pragian; lateral view.

Fig. 7. *Spyroceras latepatronus* sp. nov., MB.C.9653, holotype; bed KMO-I, Pragian; lateral view.

Fig. 8. *Spyroceras* sp. B, MB.C.9699; bed KMO-III, Pragian; lateral view.

Fig. 9. *Spyroceras cyrtopatronus* sp. nov., MB.C.9649; bed KMO-III, Pragian; lateral view.

Fig. 10. *Reticuloceras loricatum* (Barrande, 1868), MB.C.9646; bed KMO-II, Pragian; lateral view.

Fig. 11. *Spyroceras fukuijense*? Niko, 1996, MB.C.9777; bed KMO-III, Pragian; lateral view.

Fig. 12. *Probatoceras*? sp., MB.C.9651; bed KMO-II, Pragian; lateral view.

All specimens from the Filon Douze section and × 2.

PLATE 6

KRÖGER, nautiloids

The species differs from *S. patronus* in having 100, as opposed to *c.* 60, longitudinal lirae around the circumference.

Genus CANCELLSPYROCERAS gen. nov.

Type species. Orthoceras loricatum Barrande, 1868, from the Lower Devonian of Bohemia, Czech Republic.

Derivation of name. Latin, *cancellus*, lattice, referring to the net-like ornamentation of this genus, and *Spyroceras*.

Other species included. None.

Diagnosis. Very slightly cyrtoconic pseudorthocerids with reticulate ornamentation consisting of conspicuous raised longitudinal and transverse lirae; angle of expansion *c.* 11 degrees; Conch cross-section nearly circular. Approximately four septa at distance similar to conch diameter; septal necks short suborthochoanitic; siphuncle subcentral.

Remarks. Cancellspyroceras differs from *Spyroceras* in having suborthochoanitic septal necks and a shell that is not undulating. The ornamentation of *Cancellspyroceras* resembles that of *Protokionoceras* Grabau and Shimer, 1910. The internal characters of the mid Silurian type of *Protokionoceras* are poorly known; however, Sweet (1964*b*) wrote 'Siphuncle and cameral deposits in at least some species similar to those of *Geisonoceras*.' The Eifelian *Nebroceras* Zhuravleva, 1978, has a similar ornamentation but differs in having septal necks that are asymmetric along the dorsoventral axis.

Occurrence. Early Devonian; Czech Republic and Morocco.

Cancellspyroceras loricatum (Barrande, 1868)
Plate 6, figure 10; Plate 7, figure 14; Plate 8, figure 2

*1868–70 *Orthoceras loricatum* Barrande, pl. 275, figs 10–13; pl. 424, figs 5–8.
  1978 *Protobatoceras loricatum* (Barrande); Zhuravleva, p. 140.
  2001 *Protokionoceras loricatum* (Barrande); Manda, p. 6.

Material. 2 specimens, MB.C.9646, 9686, from KMO-II, Pragian, Filon Douze section.

Diagnosis. Slightly cyrtoconic with variable angle of expansion of *c.* 11 degrees; cross-section circular; surface ornamented with numerous longitudinal and directly transverse, distinctive raised lirae; longitudinal and transverse lirae form conspicuous net-like rectangular pattern;

septa directly transverse, *c.* 0.3 of cross-section diameter; septal perforation width *c.* 0.1 of cross-section diameter; septal necks short, suborthochoanitic; siphuncle subcentral on convex side of conch curvature.

Description. Larger specimen (MB.C.9646) 24 mm long, cross-section diameter 37–41 mm (angle of expansion 10 degrees). Smaller specimen (MB.C.9686) 21 mm long, cross-section diameter 23–26 mm (angle of expansion 8 degrees). Cross-section of specimens circular or slightly compressed. Ornamentation consists of equally raised transverse and longitudinal lirae (Pl. 7, fig. 10). Distance between longitudinal lirae 0.5 mm, between transverse striae 0.3–0.4 mm in MB.C.9646. Sutures straight and transverse. Distance between septa 0.3–0.4 of cross-section diameter (12 mm at cross-section diameter of 41–31 mm; 8 mm at cross-section diameter of 26 mm). Septal perforation 3 mm at cross-section diameter of 26 mm. Septal necks suborthochoanitic (Pl. 8, fig. 2).

Remarks. The type of 'Orthoceras' loricatum figured in Barrande (1868, 1870) provides no information on the position and shape of the siphuncle or of the characters of the septal necks. Therefore, to synonymize the specimens from Filon Douze with it is a little doubtful, but the fragments described are similar to 'O.' loricatum in general conch shape, ornamentation, and septal spacing. Zhuravleva (1978, p. 140) assigned 'O.' loricatum to *Probatoceras* and Manda (2001, fig. 6.) placed it in *Protokionoceras*, but neither author gave reasons for their decisions.

Cancellspyroceras loricatum differs from *Spyroceras* in having suborthochoanitic septal necks and in lacking undulation.

Occurrence. Pragian; Czech Republic and Morocco.

Genus DIAGOCERAS Flower, 1936

Type species. Orthoceras aptum Hall, 1876, from the Cherry Valley Limestone, Givetian of New York, USA.

Diagnosis (after Flower 1936, p. 293; Zhuravleva 1978, p. 127). Slender, smooth or transversely ornamented orthocones with circular or depressed cross-section; sutures oblique, sloping adapically on antisiphuncular side of conch; siphuncle subcentral; septal necks suborthochoanitic; siphuncular segments slightly expanded within chambers; mature living chamber slightly constricted; cameral deposits mural.

Remarks. The oblique septa and subcentral siphuncle are characters of *Plagiostomoceras*, which differs from *Diagoceras* in having orthochoanitic septal necks. *Diagoceras* was assigned to the Pseudorthoceratidae by Flower (1939), but its close similarity to *Adiagoceras* gen. nov.

suggests a relationship with the Arionoceratidae (see below). A definite decision with regard to the higher classification of this genus will not be possible until its apical characters are better known.

Occurrence. Eifelian–Frasnian; Morocco, North America and Russia.

Diagoceras sp.
Plate 5, figure 10

Material. 1 specimen, MB.C.9579, from bed PI, early *O. costatus* conodont Biozone, middle Eifelian, Filon Douze section.

Description. Fragment of phragmocone 62 mm long, conch height 13–28 mm (angle of expansion 14 degrees). Cross-section slightly depressed. Distance between septa 2.3 mm at apical end of fragment. Septal concavity 2.8 mm at most adapical septum. Septa oblique, sloping towards apex on antisiphuncular side. Siphuncle subcentral. Diameter of septal perforation 1.6 mm at cross-section diameter of 15 mm. Shape of siphuncular segments not preserved. Septal necks suborthochoanitic. Thin episeptal deposits occur at most apical septa.

Remarks. The closely spaced oblique septa and the suborthochoanitic septal necks allow assignment of this fragment to *Diagoceras*. It displays a significantly higher angle of expansion (15 degrees) than *Diagoceras aptum* (Hall, 1876) in which it is 6 degrees. However, the single specimen is too poorly preserved and incomplete for specific determination to be possible.

Genus SULOCERAS Manda, 2001

Type species. *Orthoceras pulchrum* Barrande, 1866, from the Konéprusy Limestone, Pragian, of Bohemia, Czech Republic.

Diagnosis (after Manda 2001, p. 277). Slender orthocones with undulating conch; cross-section circular; ornamented with alternating prominent and faint subordinate longitudinal lirae, and faint directly transverse lirae; longitudinal and transverse lirae form rectangular ornament; at ephebic stage ornamented with differentiated longitudinal striae of first to third order and faint growth lines; 3–5 septa at distance similar to cross-section diameter; sutures straight and slightly oblique; septal necks suborthochoanitic; siphuncular segments slightly expanded; mural and episeptal deposits present.

Remarks. This genus differs from *Spyroceras* in having suborthocoanitic septal necks and a conspicuous ornamentation consisting of alternating prominent and subordinate longitudinal lirae.

Occurrence. Lochkovian–Pragian; Czech Republic, Japan and Morocco.

Suloceras longipulchrum sp. nov.
Plate 7, figure 10; Plate 9, figure 5; Plate 10, figure 1

Derivation of name. Latin, *longus*, long, referring to the wide septal spacing of this species compared with other *Suloceras* species, especially the similar *S. pulchrum*.

Holotype. MB.C.9783 (Pl. 7, fig. 10; Pl. 9, fig. 5; Pl. 10, fig. 1).

Type locality and horizon. Filon Douze section, bed KMO-III, Pragian.

Material. 2 specimens, MB.C.9784–85, from the type locality and horizon.

Diagnosis. *Suloceras* with angle of expansion of 3 degrees; cross-section circular; ornamented with alternating prominent and faint subordinate longitudinal lirae; more than 60 prominent longitudinal lirae around circumference; longitudinal and transverse lirae vague; undulations shallow; *c.* 4–5 undulations at distance similar to cross-section diameter; septal distances 0.4 of cross-section diameter; sutures straight and slightly oblique, sloping in adoral direction on prosiphuncular side; septal necks suborthochoanitic; siphuncular segments nearly tubular with diameter 0.15–0.2 of cross-section; width to height ratio of siphuncular segments 0.68.

Description. Holotype 39 mm long, cross-section diameter 14–16 mm (angle of expansion 3 degrees). Cross-section circular. First order longitudinal lirae 1.3 mm apart at adoral end of holotype. Distances between directly transverse lirae *c.* 0.2 mm. Transverse and longitudinal lirae form square pattern (Pl. 7, fig. 10). Distance between shallow undulations 5 mm. Distance between two most adoral septa 6 mm. Septal perforation at most adoral septum 2.8 mm (0.18 of cross-section). Septal necks suborthochoanitic; siphuncle subcentral. Siphuncular segments slightly expanded within chambers or nearly tubular (Pl. 9, fig. 5).

Two additional specimens are poorly preserved fragments of phragmocone. MB.C.9784 has angle of expansion of 4 degrees and displays characteristic ornamentation consisting of undulations, numerous regularly spaced longitudinal lirae *c.* 0.5 mm apart, faint subordinate longitudinal lirae and faint straight, transverse lirae.

Remarks. This species differs from *S. pulchrum* in having wider septal spacing and less conspicuous, smooth undulations, and from *S. melolineatum* Niko, 1996, in having more widely spaced septa and undulations, and a wider siphuncle.

Order LITUITIDA Starobogatov, 1974
Family LAMELLORTHOCERATIADE Teichert, 1961

Remarks. Kolebaba (1999) assigned *Arthrophyllum* Beyrich, 1850 and the Lamellorthoceratidae together with *Plagiostomoceras* and *Protobactrites* to the Pallioceratida Marek, 1998, an order erected on the assumption of the presence of a cameral (pallial) mantle. This character is a physiological hypothesis. I do not consider that a taxon can be erected on such a basis, only on observed characters; hence, I do not regard the order to be valid. In fact, the different genera that have been assigned to the Pallioceratida display strong differences in conch morphology. Although *Plagiostomoceras* and *Protobactrites* are similar to lituitids in having a longitudinal lamella, which is a diagnostic character of the Lituitida, they differ significantly in septal neck shape. The longitudinal lamella is considered homoeomorphic in a variety of orthocones with wide septal spacing. Flower (1939) described a longitudinal lamella in several pseudorthoceridans. Moreover, the typical cameral deposits that cover the septal necks in apical phragmocone parts of lituitids (termed 'epichoanitic deposits' herein) do not occur in *Plagiostomoceras* and *Protobactrites*. The latter is considered an exclusive character of the Lituitida (Dzik 1984; Kröger *et al.* 2008). The presence of epichoanitic cameral deposits in *Lamellorthoceras* is the main argument for assigning this genus and the Lamellorthoceratidae to the Lituitida (Text-fig. 2). In *Sphooceras* Flower, 1962 there are deposits that are interpreted as epichoanitic (Text-fig. 16). Therefore, Dzik's (1984) opinion is followed herein and the Sphooceratidae are assigned to the Lituitida.

Genus ARTHROPHYLLUM Beyrich, 1850

Type species. *Orthoceratites crassus* Roemer, 1844, from the Wissenbach Slate, Eifelian, of Schalke, Harz Mountains, Germany.

Diagnosis (after Bandel and Stanley 1989, p. 411). Conch straight with angle of expansion of 4–12 degrees; conch cross-section circular or slightly compressed; ornamented with fine transverse ornament or with costules; siphuncle subcentral–eccentric; siphuncular segments tubular or slightly expanded within chambers; septal necks orthochoanitic; cameral deposits characterised by irregularly arranged straight to sinuous cameral lamellae, inclined longitudinally, which converge towards siphuncle.

Remarks. *Lamellorthoceras* Termier and Termier, 1950, was regarded as synonymous with *Arthrophyllum* Beyrich, 1850, by Bandel and Stanley (1989). They demonstrated that the differences in the shape of the endocameral lamella, which serve to distinguish between the two genera, are a product of fossilisation. Moreover, the conchs of *Orthoceratites crassus* and *Lamellorthoceras vermiculare* (Termier and Termier, 1950) are similarly transversely ornamented.

Occurrence. Lochkovian–Eifelian; Armorican Massif (France), Morocco, New York State and Turkey.

Arthrophyllum vermiculare (Termier and Termier, 1950)
Plate 7, figure 13; Plate 11, figures 1, 3–5; Text-figure 2

*1950 *Lamellorthoceras vermiculare* Termier and Termier, p. 41, pl. 135, figs 7–11.

EXPLANATION OF PLATE 7

Ornamentation of Early Devonian Lituitida, Orthocerida and Pseudorthocerida from the Filon Douze section.

Fig. 1. *Subdoloceras atrouzense* gen. et sp. nov., MB.C.9595, holotype; bed KMO-II, Pragian.

Fig. 2. *Subdoloceras engeseri* gen. et sp. nov., MB.C.9657; bed KMO-I, Pragian.

Fig. 3. *Spyroceras latepatronus* sp. nov., MB.C.9653, holotype; bed KMO-I, Pragian.

Fig. 4. *Spyroceras cyrtopatronus* sp. nov., MB.C.9649; bed KMO-I, Pragian.

Fig. 5. *Geidoloceras ouaoufilalense* gen. et sp. nov., MB.C.9627, holotype; bed KMO-III, Pragian.

Fig. 6. *Spyroceras patronus* (Barrande, 1866), MB.C.9652; bed KMO-I, Pragian.

Fig. 7. *Spyroceras* sp. A, MB.C.9698; bed KMO-III, Pragian.

Fig. 8. *Spyroceras* sp. B, MB.C.9699; bed KMO-III, Pragian.

Fig. 9. *Spyroceras fukuijense*? Niko, 1996, MB.C.9777; bed KMO-III, Pragian.

Fig. 10. *Suloceras longipulchrum* sp. nov., MB.C.9783, holotype; bed KMO-III, Pragian.

Fig. 11. *Anaspyroceras* sp., MB.C.9680; bed KMO-III, Pragian.

Fig. 12. *Pseudospyroceras reticulum* gen. et sp. nov., MB.C.9788, holotype; bed KMO-II, Pragian.

Fig. 13. *Arthrophyllum vermiculare* (Termier and Termier, 1950), MB.C.9648; bed KMO-V, Dalejian, Emsian.

Fig. 14. *Reticuloceras loricatum* (Barrande, 1868), MB.C.9646; bed KMO-II, Pragian.

Fig. 15. *Angeisonoceras reteornatum* gen. et sp. nov., MB.C.10092, bed KMO-II, Pragian.

All × 5.

PLATE 7

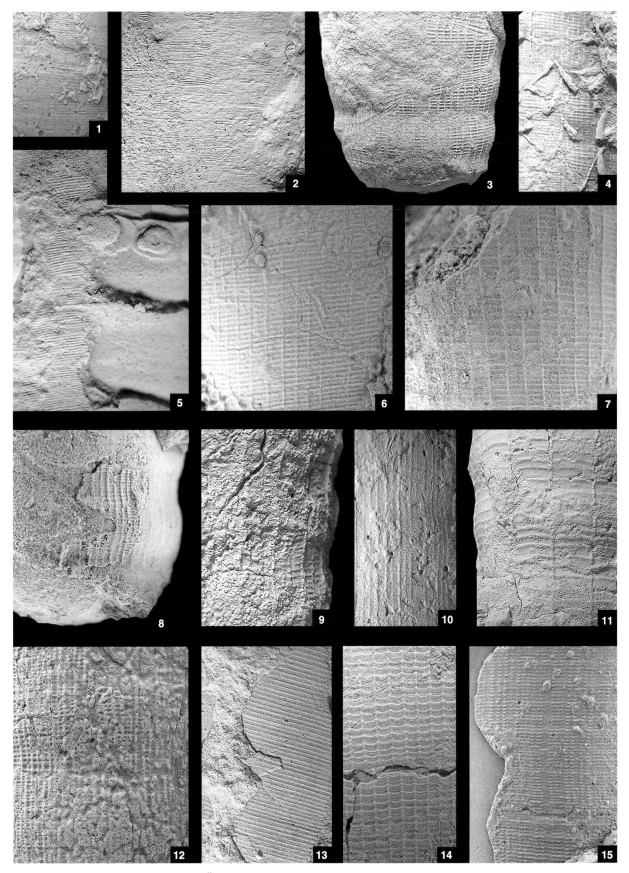

KRÖGER, Lituitida, Orthocerida, Pseudorthocerida

1950 *Lamellorthoceras gracile* Termier and Termier, p. 41, pl. 137, figs 5–6.

1966 *Arthrophyllum vermiculare* (Termier and Termier); Babin, pp. 337–340, pl. 15, figs 1–8.

1966 *Arthrophyllum gracile?* (Termier and Termier); Babin, p. 340, pl. 15, fig. 9.

1976 *Lamellorthoceras vermiculare* Termier and Termier; Stanley and Teichert, pl. 1, figs 10–11; pl. 2, figs 2–3.

1976 *Lamellorthoceras gracile* Termier and Termier; Stanley and Teichert, pl. 1, fig. 12.

1984 *Lamellorthoceras gracile* Termier and Termier; Dzik, p. 138, text-fig. 55.31.

2008 *Lamellorthoceras vermiculare* Termier and Termier; Klug *et al.*, p. 41, pl. 5, figs 3–7; pl. 6, figs 17–18.

Material. 266 specimens from beds OJ-I, MB.C.7576.1–15, KMO-I, MB.C.9578.1–10, 9578, 9693, 9752.1–11, 9753–54, 9756–57, 9789–99, KMO-II, MB.C.9781–82, 9762.1–55, KMO-III, MB.C.9525, 9765.1–13, KMO-IV, MB.C.9770.1–4, KMO-V, MB.C.9648, Erbenoceras Limestone, MB.C.9573, 9574.1–3, and F, MB.C.9572; and 6 specimens from EF, MB.C.9766.1–6; Lochkovian–Eifelian, Filon Douze section.

Emended diagnosis. Conch straight or very slightly bent, with angle of expansion below 12 degrees, most specimens with angle of expansion of *c.* 6 degrees; cross-section circular or compressed with average width to height ratio of 0.9; ornamented with fine, acutely raised lirae; lirae oblique, sloping adapically, forming broad lobe on prosiphuncular side of conch, straight on lateral sides; sutures straight; septal distance 0.2–0.45, but in most specimens *c.* 0.25 of cross-section; septal perforation subcentral; margin of septal perforation positioned at conch cross-section centre; diameter of septal perforation *c.* 0.1 of conch cross-section; siphuncular segments slightly expanded within chambers; septal necks orthochoanitic; endosiphuncular deposits not known; epichoanitic, epi-, and hyposeptal deposits present; cameral deposits form irregularly arranged radial lamellae; endocameral deposits apically cover connecting ring.

Description. Conch straight, sometimes very slightly cyrtoconic with siphuncle at convex side of curvature of growth axis. Angle of expansion between 4 and 12 degrees, average of 91 fragments 5.7 degrees (Text-fig. 14). Cross-section circular or compressed, average width to height ratio of conch cross-section of 91 fragments 0.87, standard deviation 0.08 (see Text-fig. 15). Shell ornamented with obliquely transverse, acute lirae (Pl. 7, fig. 13). Distance between lirae small, in MB.C.9648 0.3 mm at cross-section of 34 mm, in MB.C.9697 0.2 mm at cross-section of 10 mm. Ridges slope at an angle of *c.* 10 degrees towards dorsoventral axis, sloping adapically in prosiphuncular direction. Sutures straight and directly transverse at distances of 0.2 (adoral chambers of MB.C.9697) to 0.4 (adoral chambers of MB.C.9789) of conch cross-section. Most fragments have septal distances of about one-quarter of cross-section.

Largest known fragment has diameter of 44 mm (MB.C.9585) with two chambers 13 mm (adoral chamber) and 14 mm (adapical chamber) in length. MB.C.9648 slightly bent, 137 mm long, cross-section diameter of 24–41 mm. Fragment has 17 complete chambers; length of most adoral chamber 9.6 mm, most adapical 6 mm. Depth of curvature of most adapical chamber 7 mm. Septal perforation of most adapical septum 2.9 mm wide. Centre of septal perforation positioned 11.2 mm from conch margin. Septal necks orthochoanitic, slightly bent outward, deviating from direction of growth, forming slight constriction of siphuncular tube at septal perforation. Length of septal neck of second most adapical septum 0.7 mm. Siphuncular tube slightly expanded within chambers with maximum diameter close to adoral end of each segment (Pl. 11, figs 3–5). Adapical septal necks covered with 0.2-mm-thick epichoanitic deposits (Pl. 11, fig. 3). Siphuncular tube of entire specimen covered by lamellar cameral deposits. Lamellar epi- and hyposeptal deposits fill entire chambers between siphuncle and conch margin on prosiphuncular side. MB.C.9693 shows that longitudinal lamellae of cameral deposits are irregularly bent, sometimes bifurcate and merge together or are folded (Pl. 11, fig. 1).

Remarks. The species diagnosis given above is based on the original figures of Termier and Termier (1950, pl. 135, figs 7–11; pl. 137, figs 5–6) and the material described here.

Termier and Termier (1950) erected two species of *Lamellorthoceras*: a small species with a short septal distance (*L. gracile*) and a large species with a higher expansion rate and wider septal spacing (*L. vermiculare*). The material from the Filon Douze section reveals that these two species in fact represent two growth stages of the same species. The variability of conch cross-section, septal spacing and angle of expansion is considerable in the specimens measured (Text-fig. 14), but does not allow for the recognition of more species. Epichoanitic deposits in specimens MB.C.9648, 9789–99 and 9697–98 indicate that the species must be assigned to the Lituitida. *Arthrophyllum*

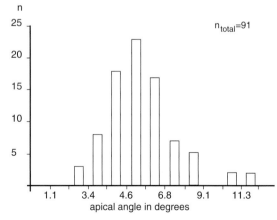

TEXT-FIG. 14. Frequency distribution of angle of expansion of 91 specimens of *Arthrophyllum vermiculare* from Filon Douze. A clear single frequency peak occurs at *c.* 5 degrees.

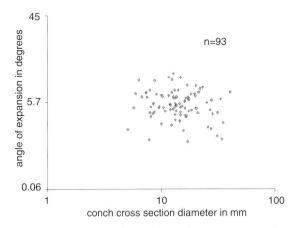

TEXT-FIG. 15. Angle of expansion and average cross section diameter of 93 fragments of *Arthrophyllum vermiculare*. No correlation is visible; axes are scaled logarithmically.

vermiculare is the most common nautiloid in bed OJ-II and the Pragian marls and limestones.

Occurrence. Lochkovian–Eifelian; France and Morocco.

Family SPHOOCERATIDAE Flower, 1962

Remarks. Dzik (1984) assigned *Sphooceras* Flower, 1962, to the Lituitida, although diagnostic characters of this group, such as longitudinal lamellae and epichoanitic deposits, had not been described in the genus. The specimens of *Sphooceras* from Filon Douze are not well preserved, but in some cases show endocameral deposits that entirely cover septal necks (Pl. 11, fig. 2; Text-fig. 16). Because of the poor preservation, the possibility that these deposits were precipitated post-mortem cannot be dismissed. The assignation of the Sphooceratidae to the Lituitida must remain questionable until better preserved material becomes available.

Genus SPHOOCERAS Flower, 1962

Type species. Orthoceras truncatum Barrande, 1860, from the upper Silurian of Bohemia, Czech Republic.

Emended diagnosis. Gradually expanding orthocones with subcentral orthochoanitic siphuncle and long chambers separated by conical septa; at apical end of all known specimens three-layered conical callus occurs, sealing siphuncular termination and smoothing contour of shell; inner layer of callus smooth; intermediate layer marked by striae that radiate outward from apex; outer layer ornamented with concentric striae broken on opposite sides of conch by lines of V-shaped invaginations in individual striae; deciduous portion of conch not known; epichoanitic and epi- and hyposeptal deposits occur.

Remarks. The generic diagnosis follows Sweet (1964*b*), who noted however: 'Neither cameral nor siphuncular deposits reported.' The specimens from Filon Douze exhibit endocameral deposits, which are, therefore, included in the diagnosis.

Occurrence. Late Silurian; Carnic Alps (Austria), Czech Republic, Morocco, Sardinia, UK and Ukraine.

Sphooceras truncatum (Barrande, 1860)
Plate 6, figure 2; Plate 11, figure 2; Text-figure 16

*1860–74 *Orthoceras truncatum* Barrande, pp. 556–559, 573–600, pl. 9, figs 1–20; pl. 342, figs 1–20; pl. 344, figs 1–6; pl. 448, figs 3–5.
1883 *Orthoceras truncatum* Barrande; Gaudry, p. 153, text-fig. 131.
1925 *Orthoceras truncatum* Barrand; Heller, p. 245, pl. 3, fig. 19.
1962 *Sphooceras truncatum* (Barrande); Flower, p. 95.
1964 *Sphooceras truncatum* (Barrande); Sweet, K231, fig. 156A5.
1975 *Sphooceras truncatum* (Barrande); Balashov, pp. 81–85, pl. 3, fig. 5.
1984 *Sphooceras truncatum* (Barrande); Dzik, pp. 112, 135, 138, pl. 31, figs 5–7, text-figs 42.19, 55.35.
1986 *Sphooceras truncatum* (Barrande); Turek and Marek, pp. 250, 252, figs 3–9.
1987 *Sphooceras truncatum* (Barrande); Kiselev *et al.*, p. 50, pl. 13, fig. 4.
1990 *Sphooceras truncatum* (Barrande); Gnoli, pp. 302, 304, pl. 4, figs 2–5.
1991 *Sphooceras truncatum* (Barrande); Gnoli and Serpagli, p. 188, 194, pl. 1, fig. 7.
1994 *Sphooceras truncatum* (Barrande); Gnoli and Kiselev, p. 416, text-fig. 1*a–c*.
1999 *Sphooceras truncatum* (Barrande); Histon and Gnoli, pp. 384, 388, tables 1–2.

Material. 5 specimens, MB.C.9702–04, 9801–02, from bed OP-M; 1 specimen, MB.C.9301, from nodules in OP-M; Ludfordian, Filon Douze section.

Diagnosis. Sphooceras with conch cross-section at position of septum of truncation of *c.* 10–20 mm; cross-section elliptically compressed; septa closely spaced, *c.* 0.3 of cross-section; sutures oblique, sloping in adapical direction on prosiphuncular side of conch; siphuncle markedly eccentric.

Description. Septum of truncation of five specimens at cross-section 10–14 mm. Cross-section elliptically compressed with width to height ratio of 0.78–0.85. Three specimens have three chambers adoral of septum of truncation; two have four chambers adoral of chamber of truncation. Septal distance invariably *c.* 0.3 of height of conch cross-section. Diameter of septal perforation 0.17 of conch height. Septal necks orthochoanitic, 0.5 mm long,

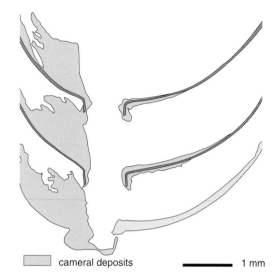

TEXT-FIG. 16. Camera lucida drawing of epichoanitic deposits in median section of *Sphooceras truncatum*; MB.C.9702, Ludlow, Filon Douze.

mature body chamber decreases towards aperture; adult cross-section diameter about 2 mm; cross-section slightly depressed or circular with flattened prosiphuncular side; ornamented with smooth undulations, sloping adapically on prosiphuncular side; undulations form conspicuous sharp lobe on prosiphuncular side and rounded saddle on antisiphuncular side; sutures straight with shallow lobes on prosiphuncular side in adolescent growth stages; at position where cross-section diameter reaches its maximum, septal distance abruptly decreases and four closely spaced sutures form wide conspicuous saddles on prosiphuncular side; most adoral of four lobate sutures less curved and compensate for following straight adoral septa; following septa nearly straight, forming faint saddle on prosiphuncular side; septal spacing in adolescent growth stages *c.* 0.4 of conch cross-section, of lobate septa 0.12–0.2 of conch cross-section, of latest septa 0.3; septal curvature shallow; septal necks orthochoanitic; septal perforations with diameter 0.16 of conch cross-section; siphuncle position eccentric; endosiphuncular and cameral deposits unknown.

one-quarter of septal perforation, one-eighth of chamber height. Most apical septa and septal necks covered with epichoanitic and hypocameral deposits that entirely cover septal necks. Deposits differ from bluish-grey sparitic chamber fillings in consisting of fine yellowish, radially arranged crystals of calcite (Text-fig. 16; Pl. 11, fig. 2).

Remarks. This species was described in detail by Barrande (1860–74) and several subsequent authors so it is not necessary to provide yet another full description. A detailed description of the cameral deposits requires better preserved material. *Sphooceras truncatum* differs from *S. amplum* Kiselev, *in* Balashov and Kiselev 1968 in being smaller at the point of truncation, and in having a more eccentrically positioned siphuncle and closer septal spacing.

Occurrence. Wenlock–Lochkovian; Carnic Alps (Austria), Czech Republic, Germany, Sardinia and UK.

Order ORTHOCERATIDA Kuhn, 1940
Family ORTHOCERATIDAE McCoy 1844

Genus CHEBBIOCERAS Klug *et al.,* 2008

Type species. Chebbioceras erfoudense Klug *et al.,* 2008, from the lower Emsian of Ouidane Chebbi, Tafilalt, Morocco.

Other species included. None.

Diagnosis. Small straight orthocones with angle of expansion in adolescent growth stages of *c.* 17 degrees; at last septum of mature specimen nearly tubular; cross-section of

Chebbioceras erfoudense Klug *et al.,* 2008
Text-figure 17A–G

    2005   *Bactrites* sp. B; Kröger *et al.,* p. 333, pl. 4C.
  * 2008   *Chebbioceras erfoudense* Klug *et al.* pp. 42–43, pl. 5, figs 16–18.

Material. 10 fragments, MB.C.9624–25, 10194, 10198–204, from bed EF, Zlíchovian, Emsian, Filon Douze section.

Diagnosis. As for genus.

Description. MB.C.10199 nearly identical to holotype (see Text-fig. 17A–B), maximum cross-section 2 mm in diameter, length 2.9 mm, five chambers preserved, similar sutural pattern and dimensions. Several specimens are fragments of phragmocone with diameter of 1.3–1.6 mm and angle of expansion of 16–17 degrees, showing characteristic ornamentation. MB.C.10198 1.0 mm long with two chambers 0.5 mm high. Adapical diameter of specimen 1.3 mm, adoral diameter 1.6 mm. Siphuncular perforation of adoral septum 0.2 mm wide. Centre of septal perforation positioned 0.5 mm from conch margin. MB.C.10198 has orthochoanitic septal necks (Text-fig. 17F). Largest specimen, MB.C.10201, with cross-section diameter of 2.2 mm and at adapical end displays one chamber 0.2 mm high. Adoral part of chamber preserved.

Genus INFUNDIBULOCERAS Klug *et al.,* 2008

Type species. Bogoslovskya mira Zhuravleva, 1978, from the Famennian of Verchi, South Ural Mountains, Russia (species combined with *Infundibuloceras* below: see 'List of new combinations').

Diagnosis. Slender, straight conch with nearly circular cross-section; surface smooth or faintly ornamented; angle of expansion below 5 degrees; sutures straight, transverse; septal distance wide, more than 0.5 of conch cross-section; siphuncle central–eccentric; septal perforation narrow, below 0.1 of cross-section; septal necks orthochoanitic and form funnel that opens gradually at adoral and adapical end; length of septal necks considerably exceeds diameter of septal perforation; no cameral or endosiphuncular deposits.

Other species included. Infundibuloceras brevimira Klug *et al.*, 2008; *I. longicameratum* sp. nov.; *I. mohamadi* sp. nov.; *Sinoceras komiense* Zhuravleva, 1978 (species combined with *Infundibuloceras* below: see 'List of new combinations').

Remarks. Infundibuloceras differs from *Bogoslovskya* Zhuravleva, 1978, in having very long septal necks that open funnel-like adorally and adapically; the septal spacing also tends to be wider. *Tibichoanoceras* gen. nov. is similar in general conch shape and also displays very long septal necks; it differs in having straight septal necks that form a strictly tubular septal perforation (Text-fig. 18D). *Infundibuloceras* differs from *Michelinoceras* in having an eccentric siphuncle (compare Text-fig. 18A, C).

Occurrence. Emsian–Famennian; Morocco and Russia.

Infundibuloceras brevimira Klug *et al.*, 2008
Plate 9, figure 11; Text-figure 18C

* 2008 *Infundibuloceras brevimira* Klug *et al.* pp. 43–44, pl. 5, figs 10–12.

Material. 5 specimens, MB.C.10071–76, from bed EF; 3 specimens, MB.C.9791–93, from bed KMO-IV, Zlíchovian, Emsian, Filon Douze section.

Diagnosis. Infundibuloceras with angle of expansion below 3 degrees; sutures straight, transverse; septal distance wide, approximately same as conch cross-section diameter; siphuncle eccentric; septal perforation narrow, 0.09 of diameter of conch cross-section; length of septal necks 0.2 of cameral length, more than three times diameter of septal perforation.

Description. Specimen MB.C.9791 76 mm long, cross-section diameter 21–24 mm (angle of expansion 2.6 degrees). Shell surface not preserved. Sutures directly transverse. Specimen has three complete septa at distances measured from adoral to adapical end of 25.5, 24.0 and 21 mm. Septal perforation positioned at most adapical septum 8 mm from conch margin. Minimum diameter of septal perforation 1.7 mm at most adoral

septum. Septal necks orthochoanitic, forming funnel that opens gradually at adoral and adapical end. Narrowest point of septal perforation at adoral third of length of septal necks. Septal neck length of most adoral septum 5 mm.

Remarks. Infundibuloceras brevimira differs from *Bogoslovskya mira* Zhuravleva, 1978 in having shorter septal necks and a lower expansion rate; the septal necks of *B. mira* are one-half of the septal distance in length (see also Klug *et al.* 2008).

Infundibuloceras longicameratum sp. nov.
Plate 12, figure 5

Derivation of name. Latin, *longus*, long, referring to the wide septal spacing of this species.

Holotype. MB.C.9705 (Pl. 12, fig. 5).

Type locality and horizon. Filon Douze section; bed PI, early *O. costatus* conodont Biozone, early Eifelian.

Other material. 1 specimen, MB.C.9696, from bed KMO-IV, Zlíchovian, Emsian, Filon Douze section.

Diagnosis. Infundibuloceras with angle of expansion below 3 degrees; sutures straight transverse; septal distance wide, *c.* 1.5 of conch cross-section; siphuncle eccentric; septal perforation narrow, 0.05 of cross-section; length of septal necks 0.2 of cameral length, approximately twice diameter of septal perforation.

Description. Holotype 79 mm long, cross-section diameter 15–19 mm (angle of expansion 2.9 degrees). Shell surface not preserved. Sutures directly transverse. Holotype has three septa at distances measured from adoral to adapical end of 30, 27 and 22 mm. Septal perforation at most adapical septum 6 mm from conch margin where conch cross-section is 16.2 mm. Minimum diameter of septal perforation 0.7 mm at most apical septum. Septal necks orthochoanitic, form funnel that opens gradually at adoral and adapical ends. Narrowest point of septal perforation at adapical third of length of septal necks. Septal neck of most apical septum 3.9 mm long.

Remarks. Infundibuloceras longicameratum differs from *Bogoslovskya mira* Zhuravleva, 1978 in having shorter septal necks and a lower expansion rate, and from *I. brevimira* in having significantly wider septal spacing and differently shaped septal necks. Its septal perforations are narrowest close to the distal tip of the septal necks, whereas those of *I. brevimira* are narrowest close to the adoral part of the necks; *Sinoceras komiense* Zhuravleva, 1978 differs in having septal necks 0.4–0.6 of the length of the chambers.

Infundibuloceras mohamadi sp. nov.
Plate 9, figure 6

Derivation of name. In honour of Mohamad, elder son of Filon Douze miner Lahcan Caraoui.

Holotype. MB.C.9687 (Pl. 9, fig. 6).

Type locality and horizon. Filon Douze section, *Erbenoceras* Limestone, Zlíchovian, Emsian.

Other material. 3 specimens, MB.C.9688–90, from the type locality and horizon.

Diagnosis. Straight, slender orthocones with angle of expansion of 4 degrees; cross-section circular; shell surface smooth; septal distance 0.6 of cross-section diameter; sutures straight and directly transverse; septal perforation between centre and conch margin; septal curvature shallow; septal necks orthochoanitic; length of septal neck *c.* 0.25 of chamber length, similar to diameter of septal perforation; diameter of septal perforation 0.1 of conch cross-section.

Description. Holotype with smooth surface, 100 mm long, circular cross-section with diameter of 15–22 mm (angle of expansion 4 degrees). Distance between two most adoral septa 10 mm, between two most adapical septa 9.6 mm. Septal concavity with depth of curvature of 6 mm at most adoral septum. Septal perforation at most adoral septum 2.2 mm (0.1 of cross-section). Septal necks orthochoanitic. Length of septal necks at most adoral septum 2.1 mm. Siphuncle position eccentric, *c.* 7 mm from conch margin at adoral end of specimen.

Additional specimens are poorly preserved phragmocone fragments. One larger specimen (MB.C.9688) displays septal spacing 0.7 of cross-section diameter at conch cross-section diameter of 15 mm. Smaller specimen (MB.C.9689) with cross-section diameter of 6–10 mm. Septal spacing of this fragment is significantly wider, being 0.8–0.9 of cross-section diameter.

Remarks. Infundibuloceras mohamadi differs from species of *Bogoslovskya* in having longer septal necks in combination with a large siphuncular diameter. It differs from *Bogoslovskya mira* Zhuravleva, 1978 in having shorter septal necks and a greater siphuncular diameter.

Genus KOPANINOCERAS Kiselev, 1969

Type species. Orthoceras jucundum Barrande, 1870, from the Ludlow of Dvoretz, Bohemia, Czech Republic.

Diagnosis (after Zhuravleva 1978, p. 53). Orthocones or very slightly cyrtocones with circular or slightly depressed cross-section; ornamented with distinct growth lines; sutures straight; siphuncle central–eccentric with cylindrical segments; septal necks funnel-like with shell thickening near distal end of necks, forming constriction of septal perforation close to tips of septal necks; endosiphuncular and cameral deposits unknown.

Remarks. Kopaninoceras differs from *Bogoslovskya* and *Infundibuloceras* in having a nearly central to slightly eccentric siphuncle; the septal necks are unique in shape,

---

EXPLANATION OF PLATE 8

Median sections of Pseudorthocerida from the Filon Douze section.

Fig. 1. *Subdoloceras atrouzense* gen. et sp. nov., MB.C.9595, holotype; bed KMO-II, Pragian; detail of septal perforation showing cyrtochoanitic septal necks.

Fig. 2. *Reticuloceras loricatum* (Barrande, 1868), MB.C.9686; bed KMO-II, Pragian; fragment with suborthochoanitic septal necks.

Fig. 3. *Subdoloceras tafilaltense* gen. et sp. nov., MB.C.9695, holotype; bed KMO-IV, Daleijan, Emsian; note the suborthochoanitic septal necks and the slightly expanded siphuncular segments.

Fig. 4. *Probatoceras?* sp., MB.C.9651; bed KMO-II, Pragian; note the angular suborthochoanitic septal necks.

Fig. 5. *Geidoloceras ouaoufilalense* gen. et sp. nov., MB.C.9627, holotype; bed KMO-III, Pragian; detail of septal perforation showing cyrtochoanitic septal necks and expanded siphuncular segments.

Fig. 6. *Subormoceras erfoudense* gen. et sp. nov., MB.C.9686, holotype; *Erbenoceras* Limestone, Zlíchovian, Emsian; detail of septal perforation; note the rounded suborthochoanitic septal necks.

Fig. 7. *Subdoloceras engeseri* gen. et sp. nov., MB.C.9654, holotype; bed KMO-I, Pragian; note the suborthochoanitic septal necks.

Figs 8, 11. *Spyroceras cyrtopatronus* sp. nov., MB.C.9676, holotype; bed KMO-III, Pragian. 8, section showing cameral deposits and cyrtochoanitic septal necks. 11, detail of septal perforation showing cyrtochoanitic septal necks and expanded siphuncular segments.

Fig. 9. *Subormoceras rissaniense* gen. et sp. nov., MB.C.9678, holotype; bed F, *P. partitus* conodont Biozone, Eifelian; section showing angular suborthochoanitic septal necks and elliptically expanded siphuncular segments. Scale bar represents 10 mm.

Fig. 10. *Subormoceras* sp., MB.C.9674; bed OSP-I, Přidolian, detail of the septal perforation with suborthochoanitic septal necks.

Scale bars represent 1 mm except where indicated.

PLATE 8

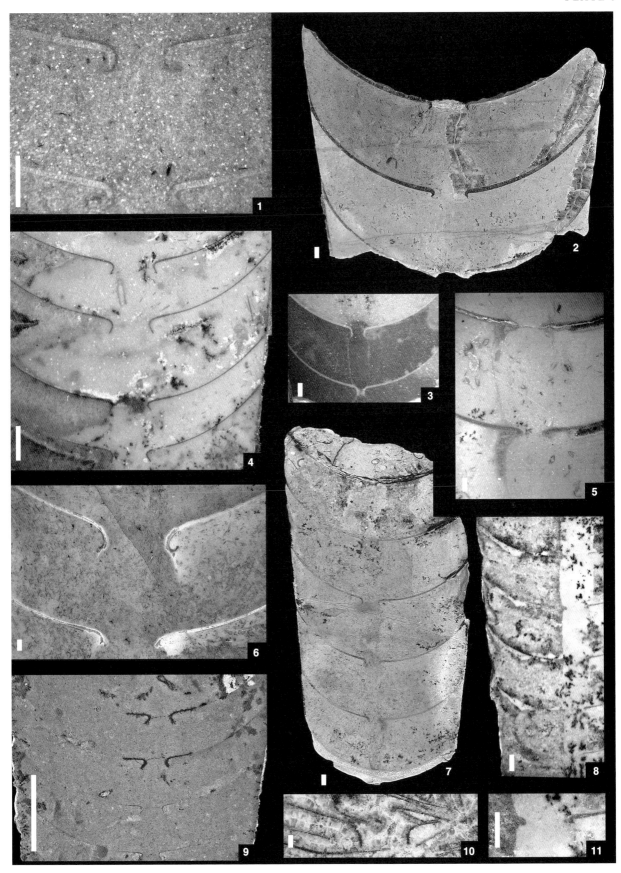

KRÖGER, Pseudorthocerida

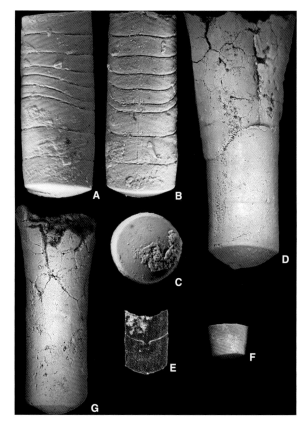

TEXT-FIG. 17. *Chebbioceras erfoudense* from the early Zlíchovian of Filon Douze. A–C, PIMUZ 7271, holotype, showing characteristic lobe pattern (compare Klug *et al.* 2008). A, lateral view. B, prosiphuncular view, showing characteristic angular lobe of ornamentation. C, adapical view. D, MB.C.10201, adult body chamber. E, MB.C.10200, adult body chamber, ventral view. F, specimen MB.C.10198, median section, showing orthochoanitic septal necks. G, MB.C.10194, fragment of two juvenile chambers, lateral view; note the large angle of expansion. All specimens × 15.

forming a funnel with a constriction close to the distal (adapical) tip of the neck (Text-fig. 18B).

Occurrence. Ludlow–Early Devonian; Carnic Alps (Austria), Czech Republic, France, Morocco, Sardinia and Ural Mountains (Russia).

Kopaninoceras? jucundum (Barrande, 1870)
Plate 9, figure 7, Plate 15, figure 5; Text-figure 18B

*1870 *Orthoceras jucundum* Barrande, p. 518, pl. 380, figs 4–7; pl. 409, figs 3–10.

1970 *Kopaninoceras jucundum* (Barrande); Barskov and Kiselev, p. 67, pl. 3, fig. 2.

1973 *Kopaninoceras jucundum* (Barrande); Kiselev, p. 125, fig. 1a–c.

1975 *Kopaninoceras jucundum* (Barrande); Chen, p. 282, pl. 4, figs 3–4, 7.

1977 *Kopaninoceras jucundum* (Barrande); Serpagli and Gnoli, p. 160, pl. 1, fig. 1a–b, text-fig. 3.

1982 *Kopaninoceras jucundum* (Barande); Kiselev and Evseev, p. 110.

1984 *Kopaninoceras jucundum* (Barande); Dzik, pp. 97, 105, pl. 25, fig. 10, text-fig. 39.23.

1999 *Kopaninoceras jucundum* (Barande); Histon and Gnoli, pp. 384, 388, tables 1–2.

Material. 2 specimens, MB.C.9526, 10085, from bed OJ-I, Lochkovian, Filon Douze section.

Diagnosis (after Serpagli and Gnoli 1977, p. 160). *Kopaninoceras* with circular cross-section; angle of expansion <8 degrees; ornamented with fine transverse lirae, slightly oblique towards growth axis, approximately ten lirae per 1 mm; septal spacing 0.6 of conch cross-section; concavity of septa about one-third of chamber length; siphuncle central; septal perforation about one-tenth of conch cross-section; septal necks orthochoanitic with length slightly more than diameter of septal perforation; siphuncular segments tubular.

Description. MB.C.9526 11 mm long, circular cross-section with diameter of 19–21 mm, angle of expansion 11 degrees. Two complete chambers preserved; length of adapical chamber 9.3 mm, of adoral chamber 10.7 mm. Septal perforation of adapical septum 1.4–1.8 mm wide. Septal necks orthochoanitic, c. 2 mm long. Septal perforation central.

MB. C.10085 17.6 mm long, minimum diameter 15 mm, angle of expansion 8 degrees. Two complete chambers preserved; length of adapical chamber 8.6 mm. Septal perforation of adapical septum c. 1.4 mm in diameter. Septal necks orthochoanitic, length c. 1.8 mm, have shell thickening and minimum diameter of central septal perforation close to adapical tip of neck (Pl. 9, fig. 7).

Remarks. The fragments described above are questionably assigned to *K. jucundum* because the diagnostically important characters of the outer shell are not preserved. The angle of expansion of both fragments is higher than that known from the type material; however, the angle of expansion in *K. thyrsus* (Barrande, 1870) varies considerably between 4 and 11 degrees (see below) which, if a similar variability for *K. jucundum* is assumed, the fragments may reflect intraspecific variation. However, more material is needed to evaluate the variability of this species.

Occurrence. Lower Silurian–Lochkovian?; Czech Republic, Morocco, Northern Urals (Russia) and Sardinia.

Kopaninoceras? dorsatoides sp. nov.
Plate 9, figure 10

Derivation of name. Latin, *dorsum*, back, lower or outer side, referring to the similarity of this species with *Kopaninoceras dorsatum* (Barrande, 1868).

Holotype. MB.C.9700 (Pl. 9, fig. 10).

Type locality and horizon. Filon Douze section, bed OP-M, Ludfordian.

Other material. 77 fragments, MB.C.9750.1–31, 9705.1–43, from the type locality and horizon.

Diagnosis. Kopaninoceras? with ovate, compressed cross-section; ratio of cross-section width to height 0.88; angle of expansion *c.* 5 degrees; concavity of septa approximately one-third of chamber length; siphuncle subcentral; diameter of septal perforation about one-tenth of conch cross-section; septal necks orthochoanitic, length about 1.5 diameter of septal perforation, or less.

Description. Holotype 41 mm long, cross-section diameter 9–13.5 mm (angle of expansion 6.3 degrees). Conch cross-section ovate with antisiphuncular side slightly narrower than prosiphuncular side. Average of angle of expansion of 78 specimens 4.6 degrees (range 1–10 degrees, standard deviation 2.3 degrees). Septal spacing *c.* 1.2 of conch cross-section in juvenile growth stages and 0.6 in later growth stages, can vary strongly within one specimen; in holotype from most adoral to most adapical chamber: 13.2, 10.8, 7.5 and 9.9 mm. Average septal spacing *c.* 0.8 of conch cross-section.

Curvature of most adoral septum of holotype 3.6 mm deep. Septal perforation of most adoral septum 6.3 mm from conch margin; of second septum from adoral end 0.8–1 mm in diameter. Septal necks orthochoanitic, 1.6 mm long, form funnel with minimum diameter of septal perforation close to adapical tip of neck.

Remarks. The fragments described above are questionably assigned to *Kopanicoceras* because the diagnostically important characters of the outer shell are not preserved. *Kopaninoceras? dorsatoides* is similar to *K. dorsatum* in having a compressed ovate cross-section and an eccentric siphuncle, but differs in having a lower expansion rate and shorter septal necks.

Kopaninoceras thyrsus (Barrande, 1870)
Plate 9, figure 9

  * 1870 *Orthoceras thyrsus* Barrande, p. 555, pl. 405, figs 15–18.

non 1960 *Orthoceras? thyrsus* Barrande; Lai, pp. 258, 269, pl. 2, fig. 7a–c.

  1968 *Michelinoceras* sp. A; Ristedt; pl. 1, fig. 4a [*fide* Kiselev 1972].

  1972 *Michelinoceras thyrsus* (Barrande); Kiselev, p. 427.

  1977 *Kopaninoceras? thyrsus* (Barrande); Serpagli and Gnoli, p. 161, pl. 1, fig. 4a–b.

  1979 *Kopaninoceras? thyrsus* (Barrande); Gnoli *et al.*, p. 418, table 4.

  1982 *Michelinoceras thyrsus* (Barrande); Kiselev and Evseev, p. 110.

  1991 *Kopaninoceras? thyrsus* (Barrande); Gnoli and Serpagli, pp. 190, 194, pl. 3, figs 4–5.

  1999 *Kopaninoceras thyrsus* (Barrande); Histon and Gnoli, pp. 384, 388, tables 1–2, fig. 4C.

Material. 80 fragments, MB.C.9540–41, 9803, 10130, 10136.1–36, 10137.1–39, 10180, from bed OP-M and limestone nodules in OP-M, Ludfordian, Ludlow; 2 specimens, MB.C.9527–9528, from bed OSP-I, Přídolí; Filon Douze section.

Diagnosis (after Serpagli and Gnoli 1977, p. 161). *Kopaninoceras* with circular cross-section; septal spacing with 1–2 chambers at distance similar to cross-section diameter; angle of expansion <4 degrees; concavity of septa approximately one-third of chamber length; siphuncle eccentric, slightly more than its own diameter removed from conch centre; diameter of septal perforation about one-tenth of conch cross-section; septal necks orthochoanitic, length about 1.5 diameter of septal perforation; thin episeptal deposits at apical parts of phragmocone.

Description. Conch circular in cross-section. Average of angle of expansion of 83 specimens 3.8 degrees (standard deviation 1.7 degrees; maximum 11 degrees, minimum 1 degree). Septal spacing can vary between 1.2 of conch cross-section diameter in juvenile growth stages and 0.4 in most adoral chambers of fragments of adult specimens. Average septal spacing *c.* 0.8 of conch cross-section.

MB.C.9803 has two chambers; length of adapical chamber 9 mm, of adoral chamber 12 mm. Conch cross-section at median septum 16 mm where septal perforation is 1.3–1.1 mm; septal necks orthochoanitic, 1.8 mm long, form funnel with minimum diameter of septal perforation close to adapical tip of neck. Centre of septal perforation of median septum 7.5 mm from conch margin. MB.C.10130 has three chambers 9.4 (adapical), 10.6 (middle), and 10.3 (adoral) mm long. Conch cross-section 16–19 mm; at adoral septum, septal perforation 1.7–1.4 mm wide; septal necks orthochoanitic, 1.8 mm long, form funnel with minimum diameter of septal perforation close to adapical tip of neck. Centre of septal perforation of adoral septum 7.5 mm from conch margin.

Remarks. Kopaninoceras thyrsus is distinguished from other species of the genus in having a combination of low expansion rate, a circular cross-section and an eccentric siphuncular position. It differs from *K. dorsatum* in having a low expansion rate and a circular cross-section, and from *K. jucundum* in having a more eccentric siphuncle and a lower conch expansion rate.

The material shows that a high variability of septal spacing exists within this species; it can vary between 0.4 and 1.2 of conch cross-section even within a single specimen. Therefore, the diagnostic value of chamber height is limited in this species.

Occurrence. Lower Silurian–Přídolí; Czech Republic, Morocco, Northern Urals (Russia) and Sardinia.

Kopaninoceras? sp.
Plate 13, figures 11–12

Material. 3 apical fragments, MB.C.10079–81, from limestone nodules in bed OP-M, Ludfordian, Filon Douze section.

Description. MB.C.10080 (Pl. 13, fig. 11) is fragment of apical 6 mm. Conch straight with circular cross-section and smooth surface. Tip blunt, after first 0.1 mm widening with angle of expansion of 4 degrees up to 1.9 mm from tip, reaching cross-section diameter of 1 mm. After 1.9 mm conch width slowly decreases for next 2 mm towards cross-section diameter of 0.9 mm when it again increases at a very low angle of expansion. Initial septa are 0.5, 0.5, 0.4, 0.5, 0.4, 0.5, 0.5 and 0.6 mm long from apex to adoral end of fragment. Maximum conch width of apex occurs at fifth chamber. Fifth to seventh septa oblique, forming conspicuous saddle on pro- or antisiphuncular side. Siphuncle eccentric.

Apical fragments MB.C.10079 (Pl. 13, fig. 12) and 10081 are similar in general shape with a straight, blunt apical angle of expansion of 4 degrees after initial 0.1 mm, decreasing conch width between 2 and 4 mm from tip, and subsequent increase with very low expansion rate. Initial septa of MB.C.10081 are 0.5, 0.25, 0.25, 0.4, 0.3, 0.4 and 0.5 mm long. Conch width in all fragments 1 mm at 5 mm from tip.

Remarks. The apices are questionably assigned to *Kopaninoceras* because they display medium septal spacing, a subcentral tubular siphuncle and a general conch shape similar to that of *Michelinoceras* sp. A. of Ristedt (1968, pl. 1, figs 3–4a), which was synonymized with *K. thyrsus* by Gnoli and Serpagli (1977). The fragments differ in having a cross-section diameter that is twice that of Ristedt's species. In bed OP-M fragments of *Kopaninoceras* several centimetres long are most common beneath apical fragments of *Arionoceras*, which are also very common in this bed. Hence, although the apices described probably represent *Kopaninoceras*, a definite generic and specific assignation will be possible only when more complete conch fragments are found.

Genus MEROCYCLOCERAS Ristedt, 1968

Type species. *M. declivis* Ristedt, 1968, from the Kok Limestone, early Ludlow, of the Plöcken area, Carnic Alps, Austria.

Diagnosis (after Ristedt 1968). Slender, straight Orthoceratidae with slightly compressed cross-section; angle of expansion *c.* 3–5 degrees; with acute, obliquely transverse undulations, sloping adapically, forming wide lobe on prosiphuncular side; undulation more pronounced in early growth stages; sutures straight; septal distance about one-third of cross-section; septal necks orthochoanitic; septal perforation subcentral to eccentric; siphuncular segments tubular; diameter of siphuncular tube about one-seventh of conch cross-section; no endosiphuncular or cameral deposits; apex with swollen first chambers.

EXPLANATION OF PLATE 9

Median sections of Orthocerida and Pseudorthocerida from the Filon Douze section.

Figs 1–2. *Spyroceras latepatronus* sp. nov., bed KMO-I, Pragian. 1, MB.C.9779 section shows short suborthochoanitic septal necks and position and size of siphuncle. 2, MB.C.9694; detail of septal perforation and siphuncular segments.

Fig. 3. *Spyroceras patronus* (Barrande, 1866), MB.C.9778; bed KMO-II, Pragian; section shows short suborthochoanitic septal necks and position and size of siphuncle.

Fig. 4. *Spyroceras fukuijense*? Niko, 1996, MB.C.9777; bed KMO-III, Pragian; note the short suborthochoanitic septal necks.

Fig. 5. *Suloceras longipulchrum* sp. nov., MB.C.9783, holotype; bed KMO-III, Pragian; detail of septal perforation and siphuncular segment.

Fig. 6. *Infundibuloceras mohamadi* sp. nov., MB.C.9687, holotype; *Erbenoceras* Limestone, Zlíchovian, Emsian; natural median section showing shape and position of septal perforation.

Fig. 7. *Kopaninoceras*? *jucundum* (Barrande, 1870), MB.C.10085; bed OJ-I, Lochkovian; detail of septal perforation and conspicuous shape of septal necks.

Fig. 8. *Orthorizoceras desertum* gen. et sp. nov., MB.C.9633; bed OSP-II, Přídolí; detail of ventral side of siphuncular segments showing two cyrtochoanitic septal necks and expanded connecting ring.

Fig. 9. *Kopaninoceras thyrsus* (Barrande, 1870), MB.C.10180; bed OP-M, Ludfordian, Ludlow; detail of septal perforation and siphuncular segment.

Fig. 10. *Kopaninoceras*? *dorsatoides* sp. nov., MB.C.9700, holotype; bed OP-M, Ludfordian, Ludlow; section shows conspicuous orthochoanitic septal necks and subcentral tubular siphuncular segments. Scale bar represents 10 mm.

Fig. 11. *Infundibuloceras brevimira* gen. et sp. nov., MB.C.9791; bed KMO-IV, Zlíchovian, Emsian. Scale bar represents 10 mm.

Scale bars represent 1 mm except where indicated.

PLATE 9

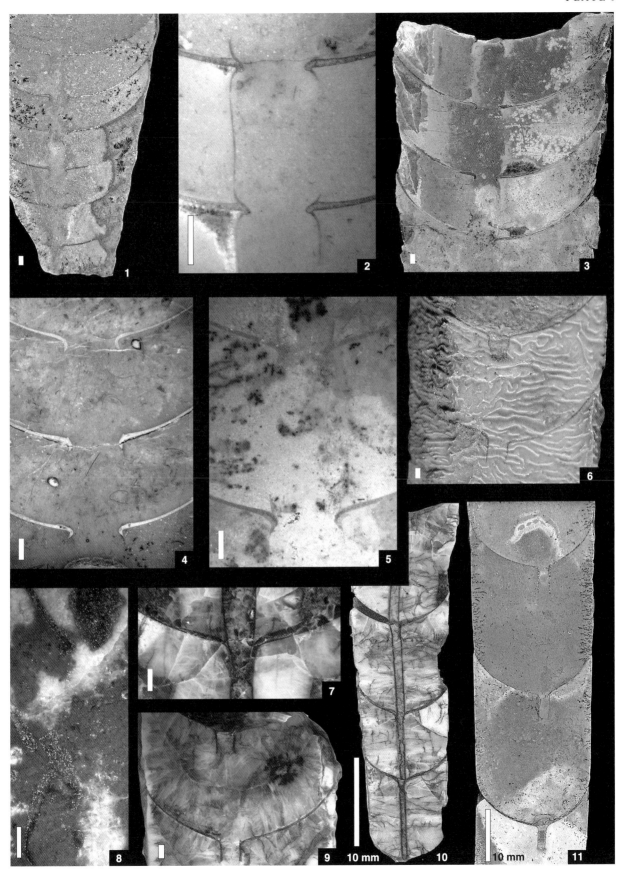

KRÖGER, Orthocerida, Pseudorthocerida

Remarks. Ristedt (1968, p. 247) provided only a very short, imprecise original diagnosis. Here I add essential information, such as the form of undulations, shape of septal necks and position of the septal perforation, based on the description of the type species provided by Ristedt.

Fragments of *Merocycloceras* that lack the apical part cannot be distinguished from *Plagiostomoceras*; both have a strongly oblique ornamentation and aperture, a subcentral tubular siphuncle, orthochoanitic septal necks, and a slender shell with slightly compressed cross-section. Therefore, it is possible that several species currently assigned to *Plagiostomoceras* will prove to be attributable to *Merocycloceras* when the apical characters are known.

Occurrence. Wenlock–Pragian; Carnic Alps (Austria), Morocco and Sardinia.

Merocycloceras sp. A
Plate 13, figure 2

Material. 1 apical fragment, MB.C.10179, from bed OJ-I, Lochkovian, Filon Douze section.

Description. Fragment of apical 4 mm; maximum diameter 0.7 mm at adoral end; ornamented with oblique, acute undulation. Angle of undulation initially strongly oblique, sloping adapically on prosiphuncular side nearly parallel to angle of expansion of initial chamber; towards adoral end decreases towards angles, departing 20–30 degrees from growth direction. Distance of annuli *c.* 0.1 mm. Apically, distance of annuli slightly wider than at adoral end of specimen. Conch cross-section circular or slightly compressed at adoral end. Apical end of initial chamber strongly depressed, forms flat, roof-shaped tip. Tip slightly bent. Initial chamber 0.5 mm long. Shallow, wide constriction occurs at 1.8 mm from tip where cross-section diameter is 0.6 mm. Maximum diameter of fragment 1.3 mm from tip where cross-section diameter is 0.7 mm. Septal distances between first four chambers are 0.5, 0.2, 0.2 and 0.3 mm measured from tip to adoral end.

Remarks. The variability of the shape of the apex and ornamentation pattern is relatively high in the specimens of *Merocycloceras* figured by Ristedt (1968, pl. 1, figs 8–11). The apex of MB.C.10179 differs from the specimens described by Ristedt in having a strongly oblique initial ornamentation and a flattened initial chamber.

Merocycloceras sp. B
Plate 13, figure 3

Material. 4 apical fragments, MB.C.9787, 9795–97, from bed KMO-I, Pragian, Filon Douze section.

Description. MB.C.9787 is fragment of apical 3 mm with maximum diameter of 1.3 mm at adoral end, ornamented with oblique, acute undulation that departs at an angle of 15–20 degrees from direction of growth throughout entire fragment. Distances between annuli *c.* 0.1 mm. Apically, distances slightly greater than at adoral end of specimen where conch cross-section is circular or slightly compressed. Apical end of initial chamber blunt and undulation appears less conspicuous. Tip slightly bent. Initial chamber slightly depressed, 0.5 mm long, with angle of expansion lower than in subsequent growth stages. Maximum diameter of fragment is at 1.3 mm from tip where width is 0.7 mm.

The three additional fragments known from bed KMO-I display apical and second chambers only. The length and maximum width of initial chamber in these specimens is 0.7 mm.

Remarks. The specimens described above differ from *M. declivis* in having a higher angle of expansion.

Merocycloceras? sp. C
Plate 13, figure 14

Material. 1 specimen, MB.C.9794, from bed OJ-I, Lochkovian, Filon Douze section.

Description. The specimen is a fragment of apical 1.7 mm with maximum diameter of 0.6 mm at adoral end. Cross-section compressed with a width to height ratio of 0.94. Ornament consists of oblique, acute undulation, which departs *c.* 15–20 degrees from direction of growth throughout entire fragment. Annuli *c.* 0.15 mm apart. Apical end of initial chamber acute and undulation appears less conspicuous. Tip slightly bent with siphuncle subcentrally positioned at convex side of growth axis. Tip cross-section grows at apical 0.8 mm towards 0.6 mm. Adapical of this position cross-section remains constant.

Remarks. The fragmentary nature of the specimen allows only a questionable assignation to *Merocycloceras*. The single definite diagnostic character visible is the oblique, acute undulation. The conch diameter does not expand adorally of 0.8 mm from tip, which is probably evidence of a slightly swollen apical interval, another diagnostic character of *Merocycloceras*. The specimen differs from *M. declivis* in having an acute tip.

Genus MICHELINOCERAS Foerste, 1932

Type species. Orthoceras michelini Barrande, 1866, from the Upper Silurian of Bohemia, Czech Republic.

Diagnosis (after Ristedt 1968, p. 53, and Kiselev 1971, p. 42). Slender orthocones with circular cross-section; angle of expansion very low, subcylindrical; shell surface smooth; sutures straight; septal distance 0.5–1 of conch cross-section; siphuncle central with diameter 0.1 of conch cross-section; siphuncular segments tubular; septal necks long, orthochoanitic, length *c.* 0.25 of chamber length, form slightly concave tube; apex displays subspherical initial chamber with shallow initial apical constriction; no conspicuous second constriction.

Remarks. Serpagli and Gnoli (1977, p.172) remarked that *Sphaerorthoceras* Ristedt, 1968, is not clearly distinct from *Michelinoceras* Foerste, 1932, and thus questioned the value of the former. Ristedt (1968, pl.1, fig. 1) illustrated an apex of *Michelinoceras michelini* (Barrande, 1866) that does not shows any apical constrictions; he also (p. 245) emended the generic diagnosis and defined *Michelinoceras* on the basis of its long, slightly concave septal necks and an apex that lacks conspicuous constriction. However, Kiselev (1971, pl. 1. fig. 3) illustrated apices of *M. michelini* that possess a shallow initial apical constriction. Thus, the shape of the apex of *M. michelini* shows some variation and the initial constriction does not appear to be a reliable character for generic determination. The main diagnostic characters of *Michelinoceras* are the septal neck shape and the very slender conch. The apically very similar *Sphaerorthoceras* differs in having shorter orthochoanitic or suborthochoanitic septal necks.

Occurrence. Late Silurian–Early Devonian?; world-wide.

Michelinoceras sp.
Plate 13, figures 16–17

Material. 2 apical fragments, MB.C.10077–78, from limestone nodules in bed OP-M, Ludfordian, Filon Douze section.

Description. MB.C.10077 (Pl. 13, fig. 17) is apical 5 mm. Conch slightly bent with circular cross-section and smooth surface. Spherical initial chamber 0.5 mm in diameter. Initial apical constriction has conch width of 0.45 mm and is 0.45 mm from tip. Conch adoral of initial constriction expands at angle of 13 degrees up to a distance of 2 mm from tip, reaching cross-section diameter of 0.75 mm. After 2 mm, conch width slowly decreases for next 1 mm, then adorally expands at very low angle. Initial septa 0.45, 0.2, 0.15, 0.15, 0.15 and 0.2 mm high, measured from apex towards adoral end of fragment.

Apical fragment MB.C.10078 (Pl. 13, fig. 16) is similar in general shape with 0.5-mm-wide spherical initial chamber and first apical constriction at 0.35 mm. Width of initial constriction 0.45 mm. Conch expands adorally of initial constriction at angle of 19 degrees up to 1.3 mm from tip where shallow second constriction occurs. Apex has angle of expansion of 4 degrees after initial 0.1 mm, decreasing conch width 2–4 mm from apex, then increasing width at very low expansion rate. Initial septa of MB.C.10078 0.35, 0.2 and 0.3 mm long. In both fragments conch width 0.5 mm at 5 mm from tip.

Remarks. The shape of the initial chamber and the smooth shell surface of these fragments are diagnostic of *Michelinoceras*. They resemble *Michelinoceras* sp. 3 of Serpagli and Gnoli (1977) and *M. currens* (Barrande, 1970). Both also show only a shallow first apical constriction and a subspherical initial chamber. However, they differ in being larger than *M. currens* and smaller than *Michelinoceras* sp. 3, and the conch section adoral of the first constriction also has a greater expansion rate than that of the latter form.

Genus ORTHOCYCLOCERAS Barskov, 1972

Type species. *Orthocycloceras alayense* Barskov, 1972, from the Upper Silurian of Isfara, Kazakhstan.

Diagnosis (after Zhuravleva 1978, p. 70). Orthocones or very slightly cyrtocones with circular cross-sections; shell ornamented with straight, transverse lirae and with straight or slightly oblique undulations or faint ribs; septal spacing wide or medium; sutures straight; siphuncle central to eccentric with cylindrical segments; septal necks orthochoanitic; endosiphuncular or cameral deposits not known.

Occurrence. Middle Silurian–Early Devonian; world-wide.

Orthocycloceras? *fluminense* (Meneghini, 1857)
Plate 4, figures 8–9

  *1857 *Orthoceras (Cameroceras) fluminense* Meneghini, p. 188, pl. C, fig. 3a–c.

  1866 *Orthoceras bohemicum* Barrande, pl. 214, figs 11–13; pl. 215, figs 8–11; pl. 288, figs 1–15; pl. 289, figs 7–8; pl. 310, figs 16–19.

  1977 *Orthocycloceras? fluminense* (Meneghini); Serpagli and Gnoli, p. 178, pl. 5, fig. 2a–b, text-fig 9 [with synonymy].

  1979 *Orthocycloceras? fluminense* (Meneghini); Gnoli *et al.*, p. 418, table 4.

  1984 '*Orthoceras*' *fluminense* (Meneghini); Dzik, pp. 108, 112, pl. 30, fig. 7, text-fig. 42.11.

  1988 *Orthocycloceras* aff. *fluminense* (Meneghini); Gómez-Alba, p. 352, pl. 173, fig. 4.

  1991 *Orthocycloceras fluminense* (Meneghini); Gnoli and Serpagli, pp. 190, 194, pl. 3, fig. 2.

  1993 *Orthocycloceras fluminense* (Meneghini); Chlupáč, pl. 6, fig. 7.

  1999 *Orthocycloceras? fluminense* (Meneghini); Histon and Gnoli, p. 389, fig. 4C.

Material. 4 specimens, MB.C.9612–15, from bed OSP-II; 1 specimen (MB.C.9685) from BS, Přídolí, Filon Douze section.

Diagnosis (compiled from Serpagli and Gnoli, 1977). Orthocones with undulating shell; angle of expansion *c.* 6 degrees; undulations rounded, oblique, sloping in adapical direction on prosiphuncular side; undulations more oblique in later growth stages (departure from growth axis 100–108 degrees); approximately three undulations at distance similar to conch cross-section; undulations gradually decrease towards antisiphuncular side, more marked at later growth stages; ornamented with irregularly spaced striae or shallow ridges parallel to undulations and with fine growth lines; septa simple, concave with spacing *c.* 0.3 of conch cross-section; diameter of septal perforation 0.1 of conch cross-section; siphuncle subcentral; septal necks orthochoanitic with length *c.* 0.25 diameter of septal perforation.

Description. Largest specimen, MB.C.9612, 33 mm long, circular cross-section 29–31 mm in diameter (angle of expansion 3.5 degrees). Undulations at distance of 10 mm oblique, sloping towards smoother conch side with departure from growth direction of *c.* 105 degrees. Maximum depth of valleys between ridges 0.8 mm. Depth gradually decreases from side where undulations form shallow saddle to opposite side where they nearly die out. Additionally ornamented with 10–11 rounded lirae per cycle of undulations (10–11 per mm) running parallel to undulations. Apical angle of other specimens varies between 7 and 9 degrees.

Remarks. Despite the large number of descriptions and illustrations in the literature, the internal characters of this species are poorly known. Unfortunately, the specimens collected from Filon Douze provides no additional information since all are heavily recrystallized. They are questionably assigned to *Orthocycloceras* because the shape of their septal necks is unknown.

Occurrence. Wenlock–Přídolí; Czech Republic, France, Morocco, Sardinia and Spain.

Orthocycloceras tafilaltense sp. nov.
Plate 10, figures 5–6; Plate 12, figure 1

2005   *Cycloceras* sp.; Kröger *et al.*, p. 331, pl. 4, figs E–F; pl. 5, fig. B.
2008   *Orthocycloceras* sp.; Klug *et al.*, pp. 44–45, pl. 5, figs 24–25, 27–30.

Derivation of name. From Tafilalt, with reference to the type region in Morocco.

Holotype. MB.C.9655 (Pl. 10, fig. 6; Pl. 12, fig. 1).

Type locality and horizon. Filon Douze section, bed KMO-II, Pragian.

Other material. 2 specimens, MB.C.9780, 10134, from the type locality and horizon; 11 specimens, MB.C.10117–10128, from bed EF, Zlíchovian, the type locality.

Diagnosis. Undulating orthocones with slightly compressed cross-section; angle of expansion 3–5 degrees;

EXPLANATION OF PLATE 10

Fig. 1. *Suloceras longipulchrum* sp. nov., MB.C.9783, holotype; bed KMO-III, Pragian; lateral view.

Figs 2–3. *Hemicosmorthoceras semimbricatum* Gnoli, 1982. 2, MB.C.9532; bed PK, Lochkovian; lateral view. 3, MB.C.10089; bed CK, Lochkovian; lateral view.

Fig. 4. *Theoceras filondouzense* gen. et sp. nov., MB.C.10095, holotype; bed F, *P. patulus* conodont Biozone, early Eifelian; lateral view.

Figs 5–6. *Orthocycloceras tafilaltense* sp. nov. 5, MB.C.10117; bed EF, Zlíchovian, Emsian; lateral view. 6, MB.C.9655, holotype; bed KMO-II, Pragian; lateral view.

Figs 7–8. *Plagiostomoceras reticulatum* sp. nov., MB.C.10116; bed OJ-I, Lochkovian. 7, lateral view. 8, adapical view.

Figs 9, 14, 18. *Plagiostomoceras culter* (Barrande, 1866). 9, MB.C.9947; bed OP-M, Ludfordian, Ludlow; detail of shell surface of body chamber showing faint oblique growth lines, lateral view. Scale bar represents 10 mm. 14, MB.C.9953; bed PK, Lochkovian; mould of body chamber, lateral view. 18, MB.C.9943; bed PK, Lochkovian; mould of body chamber and five chambers of phragmocone; lateral view.

Fig. 10. *Anaspyroceras* sp., MB.C.9680; bed KMO-III, Pragian; lateral view.

Fig. 11. *Plagiostomoceras pleurotomum* (Barrande, 1866), MB.C.9653; bed OP-K, Ludfordian, Ludlow; lateral view.

Fig. 12. *Plagiostomoceras lategruenwaldti* sp. nov., MB.C.9954; bed OJ-I, Lochkovian; lateral view.

Figs 13, 17. *Plagiostomoceras bifrons* (Barrande, 1866); bed OJ-I, Lochkovian. 13, MB.C.9944; lateral view. 17, MB.C.10088; juvenile specimen; lateral view.

Fig. 15. *Angeisonoceras reteornatum* gen. et sp. nov., MB.C.9946, holotype; bed KMO-II, Pragian; lateral view.

Fig. 16. *Pseudospyroceras reticulum* gen. et sp. nov., MB.C.9788, holotype; bed KMO-II, Pragian; lateral view.

All specimens from the Filon Douze section and × 2 except where indicated by the scale bar.

PLATE 10

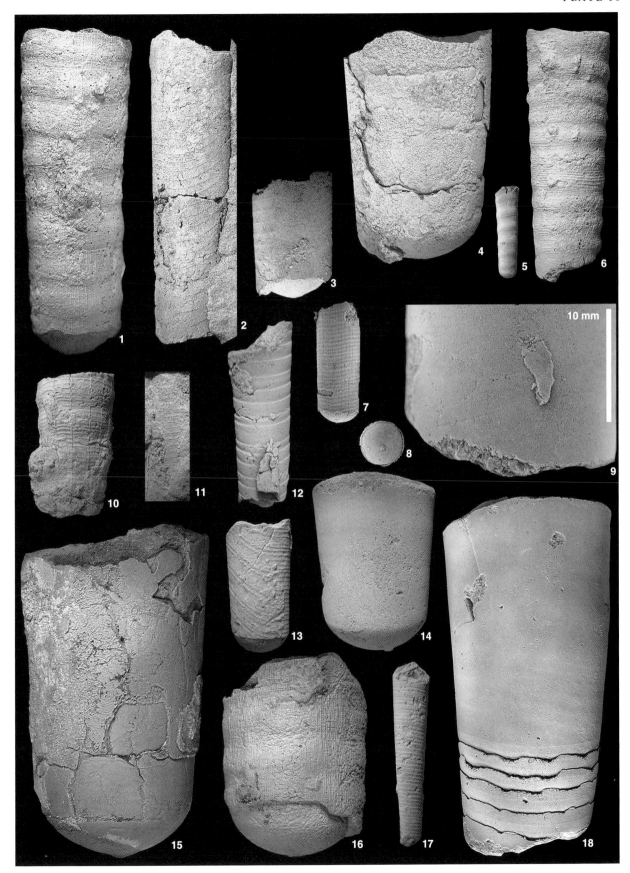

KRÖGER, nautiloids

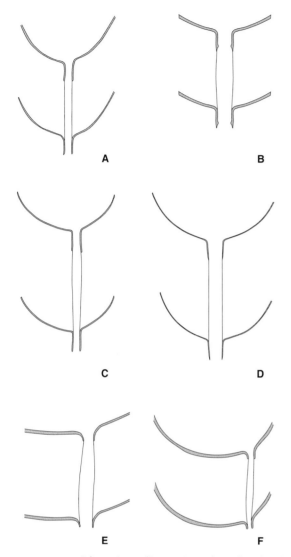

TEXT-FIG. 18. Schematic median sections of septal necks of selected Orthoceratidae. A, *Michelinoceras michelini*, after Sweet (1964*b*). B, *Kopaninoceras jucundum*; MB.C.10085. C, *Infundibuloceras brevimira*; MB.C.9791. D, *Tibichoanoceras tibichoanum* gen. et sp. nov.; MB.C.9949. E, *Theoceras filondouzense* gen. et. sp. nov.; MB.C.10095. F, *Bogoslovskya perspicua*, after Zhuravleva (1978, pl. 3, fig. 4*a*). *Theoceras* differs from *Bogoslovskya* in having a transversely ornamented shell. Not to scale.

undulations slightly oblique, sloping in adoral direction on prosiphuncular side; ornamented with rounded ridges parallel to undulations, 9–15 ridges occur per cycle of undulation; one cycle of undulation between two septa; sutures nearly straight; septal distance 0.3–0.5 of conch cross-section; siphuncle subcentral with tubular segments; diameter of septal perforation *c.* 0.1 of conch cross-section; septal necks orthochoanitic, length more than one-half of diameter of septal perforation.

Description. Holotype is fragment of phragmocone 30 mm long, cross-section circular with diameter 10–12 mm (angle of expansion 3 degrees). Slightly oblique, transverse undulations sloping in adoral direction on prosiphuncular side, forming shallow lobe on antisiphuncular side and shallow sinus on prosiphuncular side (Pl. 10, fig. 6). Distance between undulations *c.* 3 mm (four undulations occur at distance similar to conch cross-section). Additionally, narrow transverse lirae and valleys parallel to undulations occur, nine within 2 mm, *c.* 15 at distance similar to one cycle of undulations.

MB.C.10117 27 mm long, cross-section diameter 7.5–9.5 mm (angle of expansion 4.2 degrees). One cycle of undulations occurs at distance similar to chamber height. Fragment has eight complete chambers with chamber height of *c.* 2.7–3.5 mm (0.3–0.5 of conch cross-section). Siphuncle subcentral. Siphuncular segments tubular (Pl. 12, fig. 1). Septal perforation at most adapical septum 0.8 mm (*c.* 0.1 of conch cross-section). Septal necks orthochoanitic, 0.5 mm long at most adapical septum.

Smaller fragments more slender with greater septal distance and more distance between undulations. MB.C10117 (Pl. 10, fig. 5), fragment of juvenile body chamber and septum at base of chamber, length 10 mm, cross-section diameter 2.5–3.4 mm (angle of expansion 5 degrees). Undulations oblique, sloping in adoral direction on prosiphuncular side at distance of 1.4 mm. Additionally six lirae occur per 1 mm, running parallel to undulations (*c.* 9 per cycle of undulations).

Remarks. Orthocycloceras tafilaltense differs from juvenile parts of *Orthoceras agassizi* Barrande, 1866, in having a thin, tubular, nearly central siphuncle that lacks endosiphuncular deposits. *Orthocycloceras allapsum* Zhuravleva, 1978, differs in having a higher angle of expansion, more closely spaced transverse striations, and non-oblique undulation.

Genus PSEUDOSPYROCERAS gen. nov.

Type species. Pseudospyroceras reticulum sp. nov., from the Pragian of the Filon Douze section.

Derivation of name. Greek, *pseudes*, false, referring to the ornamentation of this genus, which is similar to that of *Spyroceras.*

Other species included. None.

Diagnosis. Slender orthoconic, undulating longicones; conch cross-section nearly circular; ornamented with narrowly spaced, raised longitudinal and transverse lirae; distance between longitudinal lirae slightly greater than between transverse lirae, forming a reticulate ornament; two undulations at distance similar to distance between two successive septa; septal spacing wide, approximately two septa at distance similar to conch diameter; septal necks orthochoanitic; siphuncle central to subcentral; siphuncular diameter 0.1 of conch cross-section.

Remarks. Although the ornamentation of *Pseudospyroceras* is similar to that of *Spyroceras*, the septal necks are orthochoanitic and the siphuncle is narrow, as in typical Orthocerida. Niko (1996, fig. 4.10) described spyroceratids with suborthochoanitic to orthochoanitic septal necks in ephebic growth stages; these differ from *Pseudospyroceras* in having wider and more expanded siphuncular segments and septal necks that are slightly bent outwards. However, *P. reticulatum* is based on a single, relatively large fragment; additional material may indicate that it is from an adult *Spyroceras*. *Pseudospyroceras* is probably related to *Orthocycloceras* or *Kionoceras* Hyatt, 1884.

Occurrence. Pragian; Morocco.

Pseudospyroceras reticulum sp. nov.
Plate 7, figure 12; Plate 10, figure 16; Plate 12, figure 2

Derivation of name. Latin, *reticulum*, net, referring to the ornamentation of the species.

Holotype. MB.C.9788 (Pl. 7, fig. 12; Pl. 10, fig. 16; Pl. 12, fig. 2).

Type locality and horizon. Filon Douze section, bed KMO-II, Pragian.

Other material. Known only from the holotype.

Diagnosis. As for genus.

Description. Holotype is fragment of phragmocone 17 mm long, cross-section circular with diameter 19–20 mm (angle of expansion 6 degrees). Ornamented with four straight, transverse undulations 5 mm apart, longitudinal lirae 0.3 mm apart, transverse lirae 0.1–0.2 mm apart. Some longitudinal lirae more distinct than others, generally more prominent than transverse ones. Three septa of holotype 7.5 and 9 mm apart. Diameter of perforation of adoral septum 1.7 mm. Septal necks orthochoanitic (Pl. 12, fig. 2).

Remarks. This species, known only from a single fragment, displays all the diagnostic characters that are needed to distinguish it; the combination of orthochoanitic septal necks, an undulating shell and a reticulate ornamentation is unique. It is assigned to the Orthoceratidae on the basis of the first of these characters.

Genus THEOCERAS gen. nov.

Type species. *Theoceras filondouzense* sp. nov. from the Eifelian of the Filon Douze section.

Derivation of name. In honour of Theo Engeser, Berlin, who has devoted much of the last decade to compiling a taxonomic database of fossil non-ammonoid cephalopods.

Other species included. None.

Diagnosis. Slender, straight conch with circular cross-section; surface ornamented with minute transverse striae, which form shallow lateral lobes; angle of expansion *c.* 8–11 degrees; sutures straight, transverse or slightly oblique with antisiphuncular side sloping towards apex; septal distance 0.4 of conch cross-section diameter; siphuncle eccentric; siphuncular segments tubular; siphuncular diameter *c.* 0.05 of cross-section; septal necks orthochoanitic, form funnel with smallest diameter at adapical tip of necks; no cameral or endosiphuncular deposits.

Remarks. *Theoceras* is similar to *Bogoslovskya* Zhuravleva, 1978, in having a slender conch and a very narrow, eccentric siphuncle (compare Text-fig. 18E–F). However, *Bogoslovskya* differs in having a smooth shell and wider septal spacing.

Occurrence. Eifelian; Morocco.

Theoceras filondouzense sp. nov.
Plate 10, figure 4; Plate 12, figure 3; Text-figure 18E

Derivation of name. From Filon Douze, with reference to the type locality.

Holotype. MB.C.10095 (Pl. 10, fig. 4; Pl. 12, fig. 3; Text-fig. 18E).

Type locality and horizon. Filon Douze section, bed F, *P. patulus* conodont Biozone, early Eifelian.

Other material. 2 specimens, MB.C.10096–97, from the type locality and horizon.

Diagnosis. As for genus.

Description. Holotype 14 mm long, cross-section circular, 15–17 mm in diameter (angle of expansion 8 degrees). Ornamented with fine transverse striae that form shallow lateral lobes; distance between two striae *c.* 0.1 mm. Fragment displays three chambers. Distance between two most adoral septa 5.2 mm, between most adapical chambers 4.6 mm. Septa slightly oblique, sloping adapically on antisiphuncular side. Septal perforation at most adoral septum 6.7 mm from conch margin with diameter of *c.* 0.7 mm (0.05 of cross-section). Septal necks orthochoanitic, form funnel with smallest diameter at tips. Siphuncular segments tubular.
MB.C.10096 has poorly preserved surface; length 26 mm; cross-section circular, 17–22 mm in diameter (angle of expan-

sion 11 degrees), with four chambers. Distance between two most adoral septa 6.3 mm, between most adapical chambers 5.7 mm. Septa directly transverse. Septal perforation at most adoral septum 8.6 mm from conch margin with diameter of *c.* 1 mm (0.05 of cross-section). Septal necks orthochoanitic, form funnels with smallest diameter at tips (Pl. 12, fig. 3). Siphuncular segments tubular.

Genus TIBICHOANOCERAS gen. nov.

Type species. *Tibichoanoceras tibichoanum* sp. nov., from the Pragian of the Filon Douze section.

Derivation of name. Latin, *tibia*, flute, referring to the narrow septal perforation with very long septal necks of this genus.

Other species included. *Michelinoceras abditum* Kiselev, *in* Balashov and Kiselev 1968; *Sinoceras riphaeum* Zhuravleva, 1978 (both species combined with *Tibichoanoceras* below: see 'List of new combinations').

Diagnosis. Slender, straight conch with nearly circular cross-section; surface smooth; angle of expansion below 10 degrees; sutures straight, transverse; septal distance wide, more than 0.5 of conch cross-section; siphuncle central–eccentric; siphuncular segments tubular; siphuncular diameter *c.* 0.1 of conch cross-section; septal necks long, orthochoanitic, form straight tube; transition from septal necks to septa a sharp angle of *c.* 90 degrees; thickness of septal necks gradually decreases towards tip of necks; no cameral or endosiphuncular deposits.

Remarks. *Tibichoanoceras* differs from *Michelinoceras* Foerste, 1932, in having a wider siphuncle and in the shape of the septal necks. The septal necks of *Michelinoceras* form a narrow funnel that gradually opens towards the adoral and adapical end of the septal perforation, whereas those of *Tibichoanoceras* form a straight tube with a sharp angle at the transition between septum and septal neck (compare Text-fig. 18A, D). This septal neck shape is similar to that of *Sinoceras* Shimizu and Obata, 1935*b*, which is, however, ornamented with conspicuous

transverse, lobate striae (see Yü 1930; Lai 1986) and has endosiphuncular and cameral deposits that are characteristic for lituitids.

Occurrence. Ludfordian–Emsian; Morocco, Ukraine and Russia.

Tibichoanoceras tibichoanum sp. nov.
Plate 12, figure 6; Text-figure 18D

Derivation of name. As for genus.

Holotype. MB.C.9949 (Pl. 12, fig. 6; Text-fig. 18D).

Type locality and horizon. Filon Douze section, bed KMO-I, Pragian.

Other material. Only known from the holotype.

Diagnosis. Slender, straight conch with nearly circular cross-section; surface smooth; angle of expansion below 6 degrees; sutures straight, transverse; septal distance 0.8 of conch cross-section; siphuncle central; siphuncular segments tubular; siphuncular diameter *c.* 0.1 of conch cross-section; septal necks long, orthochoanitic, form straight tube; transition from septal necks to septa a sharp angle of *c.* 90 degrees; length of septal necks *c.* 0.2 of septal distance, 1.5 of diameter of septal perforation; thickness of septal necks gradually decreases towards tip of necks.

Description. Holotype with smooth surface, 70 mm long, cross-section circular, diameter 40–47 mm (angle of expansion 6 degrees). Fragment displays two complete chambers. Distance between two most adoral septa 37 mm, between most adapical chambers 32 mm. Septal curvature with depth of 18 mm at most adoral septum. Sutures straight, transverse. Septal perforation at most adoral septum 5 mm in cross-section. Septal necks long, orthochoanitic and straight. Transition from septal necks to septa a sharp angle of *c.* 90 degrees. Length of septal necks 7.5 mm at most adoral septum, *c.* 0.2 of septal spacing, 1.5 of diameter of septal perforation. Siphuncle central. Siphuncular segments tubular.

EXPLANATION OF PLATE 11

Lituitida from the Filon Douze section.

Figs 1, 3–5. *Arthrophyllum vermiculare* (Termier and Termier, 1950). 1, MB.C.9693; bed KMO-I, Pragian; weathered surface of lamellar cameral deposits; × 2. 3, MB.C. 9697; bed KMO-I, Pragian; median section of apical conch section showing epichoanitic deposits and lamellar deposits that partially cover the connecting ring which is not preserved. 4–5, MB.C.9648; bed KMO-V, Dalejian, Emsian; median section showing expansion of lamellar cameral deposits and epichoanitic deposits in apical part.

Fig. 2. *Sphooceras truncatum* (Barrande, 1860), MB.C.9702; bed OP-M, Ludfordian, Ludlow; median section showing detail of septal perforation and siphuncular segments; note the epichoanitic deposits.

Scale bars represent 1 mm.

PLATE 11

KRÖGER, Lituitida

Remarks. This species differs from *Michelinoceras abditum* Kiselev, *in* Balashov and Kiselev 1968 in having a narrower siphuncule and longer septal necks, and from *Sinoceras riphaeum* Zhuravleva, 1978 in having wider septal spacing and a central siphuncle.

Orthoceratidae gen. et sp. indet.
Plate 16, figure 3

Material. 1 specimen, MB.C.9913, from KMO-V, Daleijan, Emsian, Filon Douze section.

Description. Fragment is complete body chamber with four chambers of phragmocone; total length 39 mm; conch cross-section nearly circular with adapical diameter of 9 mm, adoral diameter 11 mm. Length of body chamber *c.* 22 mm. Maximum cross-section diameter of specimen 12 mm and positioned at mid-length of body chamber. Aperture slightly constricted. Conch slightly cyrtoconic with siphuncle eccentrically positioned on convex side of curvature of growth axis. Septal distances 4.3, 5.0, 5.6 and 3.5 mm, respectively, measured from adapical end. Septal concavity relatively shallow with a depth of curvature of 2 mm at most apical septum. Width of septal perforation 0.7 mm at conch cross-section of 10 mm in diameter where length of orthochoanitic septal neck *c.* 0.8 mm. Siphuncular segments tubular. No endosiphuncular or cameral deposits.

Remarks. The fragment belongs to a mature individual, as indicated by the shape of the body chamber, which is constricted at the aperture, and by the crowded adoral septa. The specimen represents a small orthocerid that is comparable in general form with *Michelinoceras? sulciferum* Zhuravleva, 1978, and *Mericoceras* Zhuravleva, 1978. However, the surface characters of the fragment are not preserved, rendering generic and specific determination impossible.

Family ARIONOCERATIDAE Dzik, 1984

Genus ARIONOCERAS Barskov, 1966

Type species. Orthoceras canonicum Meneghini, 1843, from the Wenlock of Cea di San Antonio, Fluminimaggiore, Sardinia, Italy.

Diagnosis (after Serpagli and Gnoli 1977, p. 182). Conch straight or weakly curved, narrowly conical, moderately dilating; angle of expansion *c.* 6–10 degrees; cross-section circular to compressed; shell surface smooth or with shallow transverse sculpture; siphuncle central; septal necks suborthochoanitic, very short, acuminate, their lengths less than diameter of septal foramen; connecting ring cylindrical; protoconch large, pointed apically, ogival in shape and bent ventrally; long caecum crosses whole protoconch; cameral deposits in apical chambers of adult specimens.

Remarks. Orthoceras arion Barrande, 1866, was chosen by Barskov (1966) as type species of the genus. However, the name is a junior subjective synonym of *Orthoceras affine* Meneghini, 1857, which is preoccupied by *O. affine* Portlock, 1843. The next available valid name for *O. arion* is *Orthoceras canonicum* Meneghini, 1857 (Serpagli and Gnoli 1977; Gnoli 1998).

Serpagli and Gnoli (1977) described the cross-section of *Arionoceras* as circular. However, some variation from this occurs in the type species. Herein, I extend the diag-

EXPLANATION OF PLATE 12

Median sections of Orthocerida from the Filon Douze section.

Fig. 1. *Orthocycloceras tafilaltense* sp. nov., MB.C.9780; bed KMO-II, Pragian; note the tubular siphuncle and orthochoanitic septal necks; scale bar represents 1 mm.

Fig. 2. *Pseudospyroceras reticulum* gen. et sp. nov., MB.C.9788, holotype; bed KMO-II, Pragian; section shows position and size of septal perforation and short, orthochoanitic septal necks.

Fig. 3. *Theoceras filondouzense* gen. et sp. nov., MB.C.10095, holotype; bed KMO-III, Pragian; note the eccentric position of siphuncle and short, orthochoanitic septal necks.

Fig. 4. *Arionoceras canonicum* (Meneghini, 1857), MB.C.9879; bed OP-M, Ludfordian, Ludlow; note the very short septal necks.

Fig. 5. *Infundibuloceras longicameratum* gen et. sp. nov., MB.C.9705, holotype; bed PI, *O. costatus* conodont Biozone, Eifelian; note the extremely large septal distance.

Fig. 6. *Tibichoanoceras tibichoanum* gen. et sp. nov., MB.C.9949, holotype; bed OJ-I, Lochkovian; detail of septal perforation and conspicuous shape of septal necks.

Fig. 7. *Temperoceras? ludense* (Sowerby, *in* Murchison 1839), MB.C.9799; bed OP-M, Ludfordian, Ludlow; section shows characteristic annular deposits.

Fig. 8. *Arionoceras capillosum* (Barrande, 1867), MB.C.10205; bed BS, Lochkovian.

Scale bars represent 10 mm except where indicated.

PLATE 12

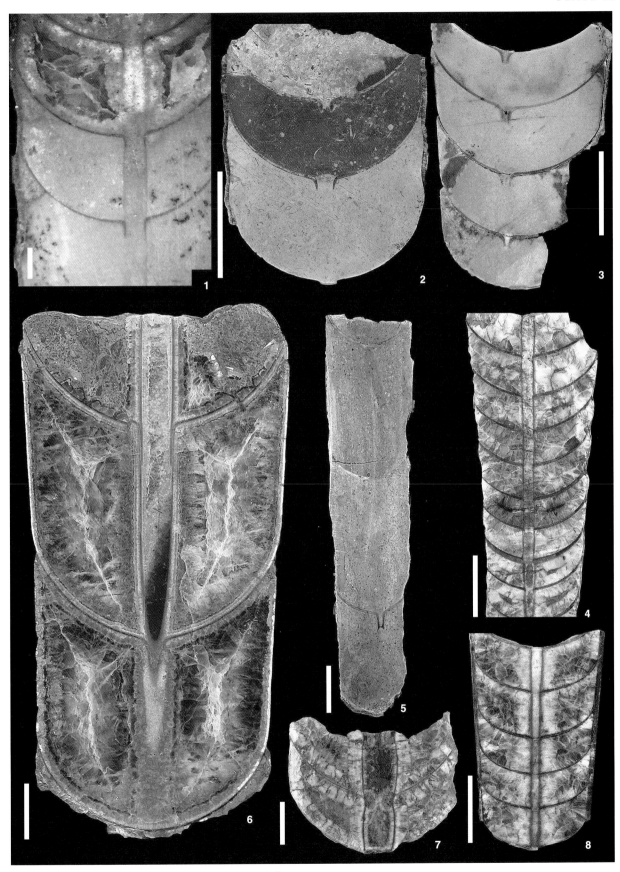

nosis to include forms with subcircular and compressed cross-sections but consider this to be a refinement of the wording of the diagnosis rather than an emendation.

The description of *Arionoceras lotskirkense* Kröger, 2004, extended the generic time range of the genus to the Middle Ordovician. However, the septal necks of the Middle Ordovician species are considerably longer than those of their Silurian counterparts and apical parts of the species are not known, making the generic assignment questionable. Hence, the generic time-range is considered here to be Silurian–Early Devonian.

Occurrence. Mid Silurian–Lochkovian; Armorican Massif (France), Carnic Alps (Austria), Czech Republic, Morocco and Sardinia.

Arionoceras canonicum (Meneghini, 1857)
Plate 12, figure 4; Plate 13, figures 23–29

```
 1857   Orthoceras affine Meneghini, p. 217, pl. C, fig. 16.
*1857   Orthoceras canonicum Meneghini, p. 199, pl. C,
          fig. 7a′–a.
 1866   Orthoceras arion Barrande, p. 221, fig. 2.
 1977   Arionoceras affine. (Meneghini); Serpagli and Gnoli,
          p. 182, pl. 6, figs 2–7, text-fig. 10b [with synonymy].
 1979   Arionoceras arion (Meneghini); Babin in Babin et al.,
          p. 73, pl. 8, figs 5–11.
 1998   Arionoceras affine (Meneghini); Gnoli, p. 2.
 1998   Arionoceras arion (Meneghini); Gnoli, p. 2.
 1999   Arionoceras affine (Meneghini); Histon and Gnoli,
          pp. 384, 388, tables 1–2.
```

Material. 142 fragments, MB.C.9800.1–141 and 9879, and 46 apices, MB.C.10231–10235, 10139.1–41, from bed OP-M, Ludfordian, Ludlow, Filon Douze section.

Diagnosis (after Serpagli and Gnoli 1977, p. 182). Arionoceratids with subcircular cross-section; angle of expansion at adapical part of conch 5–6 degrees, increasing adorally to 7–8 degrees; shell surface ornamented with fine, irregular growth lines, three per 1 mm; suture straight, transverse; chambers short *c.* 0.5 of cross-section diameter; depth of septal curvature one-third of septal distance; siphuncle subcentral with diameter *c.* 0.17 of conch cross-section; septal necks very short, suborthochoanitic–loxochoanitic; connecting rings slightly expanded within chambers; mural and episeptal deposits in apical part of adult specimens; initial chamber large, pointed apically and ventrally bent, with width to length ratio of *c.* 0.83; long ceacum crosses whole protoconch.

Description. Conch has nearly circular cross-section. Average angle of expansion of 143 specimens 8 degrees (standard deviation 2.3 degrees; maximum 15 degrees, minimum 3 degrees;

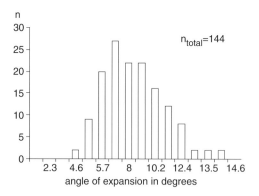

TEXT-FIG. 19. Frequency distribution of angle of expansion of 144 specimens of *Arionoceras canonicum*. A clear single frequency peak occurs at *c.* 7 degrees.

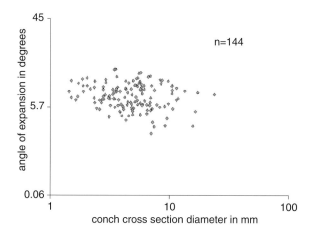

TEXT-FIG. 20. Angle of expansion and average cross-section diameter of 93 fragments of *Arionoceras canonicum*. Note the inconspicuous tendency of a lower angle of expansion in the specimens with a larger cross-section. Axes logarithmically scaled.

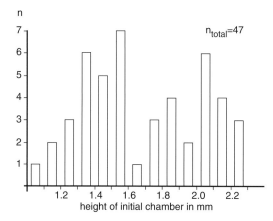

TEXT-FIG. 21. Frequency distribution of height of initial chamber of 47 specimens of *Arionoceras canonicum*. Note the two frequency peaks (compare with Serpagli and Gnoli 1977).

Text-fig. 19). No change in angle of expansion during growth visible (Text-fig. 20), but because fragments smaller than 5 mm in cross-section are usually preserved only as disarticulated chambers, estimation of expansion rate of early juvenile stages is difficult. Variability of septal spacing relatively low. Septal distance 0.3–0.5 of conch cross-section diameter. Sutures straight, transverse. Septal necks short.

MB.C.9879 62 mm long, cross-section diameter *c.* 14–24 mm, with 11 complete chambers. Distance between chambers varies between 3.8 and 6.3 mm. Length of second chamber from adoral end of fragment 6 mm, with cross-section diameter 20.6–21.5 mm. Septal perforations of two septa of this chamber 1.6 mm. Centre of septal perforation 10.3 and 9.2 mm from conch margin, respectively. Septal necks very short, *c.* 0.2 mm, loxochoanitic. Siphuncular segments 1.8 mm in diameter, essentially tubular, forming constriction at septal perforation.

Forty-seven apices with elongate, ovate initial chamber with acute tip and shallow apical constriction. Initial chamber slightly bent. Long tubular ceacum present within initial chamber. Two size classes can be distinguished (Text-fig. 21). Small apices with initial chamber length 1.3–1.5 mm (Pl. 13, figs 25–26, 28) and large apices with initial chamber length 2.0–2.2 mm (Pl. 13, figs 23–24, 27, 29). In large and small apices constriction occurs at cross-section diameter of 1.0–1.1 mm. Conch adoral of initial constriction grows at angle of expansion of 8–9 degrees. Several apices preserved as sparitic moulds of former cameral and siphuncular spaces (Pl. 13, figs 23–26).

Remarks. This species was described in great detail by Serpagli and Gnoli (1977); the description of the specimens here contribute to knowledge of the intraspecific variation. Serpagli and Gnoli (1977) did not include the term 'loxochoanitic' in their specific diagnosis. However, they emphasized the very short septal necks and their text-figure 10 indicates that the species has loxochoanitic septal necks. This is not considered to be an emendation of the diagnosis, rather a more precise statement of what Serpagli and Gnoli (1977) expressed.

Occurrence. Ludlow–Přídolí; Armorican Massif (France), Carnic Alps (Austria), Czech Republic, Morocco, Russia and Sardinia.

Arionoceras aff. *canonicum* (Meneghini, 1857)
Plate 16, figures 4, 7

Material. 65 fragments, MB.C. 9751.1–63, 10229-10230, from bed OP-M, Ludfordian, Filon Douze section.

Description. MB.C.10229 34 mm long, conch cross-section elliptically compressed with width to height ratio of *c.* 0.9; width 22–28 mm, height 25–30 mm (angle of expansion 8.4 degrees). Specimen has four complete chambers; chamber length 6–7 mm. Sutures transverse with shallow lateral lobes. Septal curvature of most adoral septum 7 mm. Diameter of septal perforation at most adoral septum 2.4 mm, which is 0.8 of conch cross-section.

Centre of septal perforation at most adoral septum 14.4 mm from conch margin. Septal necks very short, suborthochoanitic. Length of septal neck of most adoral septum 0.3 mm. Siphuncular segments slightly expanded within chambers, elliptical in median section, forming constriction at septal perforation. Maximum diameter of siphuncular segment at most adoral chamber 2.8 mm. Average angle of expansion of all 65 specimens 6.3 degrees (standard deviation 2.3 degrees, maximum 12 degrees, minimum 1 degree). Width to height ratio of 14 representative specimens averages 0.91, maximum 0.96, minimum 0.84.

Remarks. These specimens are distinguished from the type of *A. canonicum* in having a clearly compressed cross-section. However, clear discrimination of a species with a compressed cross-section and *A. canonicum* with a subcircular cross-section is impossible because all stages between circular and compressed forms exist at Filon Douze.

Arionoceras capillosum (Barrande, 1867)
Plate 12, fig 8; Plate 16, figures 5–6

  *1868 *Orthoceras capillosum* Barrande, pl. 325, figs 19–33; pl. 357, figs 4–7; pl. 394, figs 16–19.

   1881 *Orthoceras capillosum* Barrande; Maurer, p. 27, pl. 2, fig. 5.

   1890 *Orthoceras capillosum* Barrande; Whidborne, pp. 129, 131.

   1914 *Orthoceras capillosum* Barrande; Chapman, pp. 206, 216.

   1950 *Orthoceras* cf. *capillosum* Barrande; Termier and Termier, p. 36, pl. 132, figs 24–25.

   1972 *Michelinoceras capillosum* (Barrande); Barskov, p. 37, pl. 1, figs 3–4; pl. 2, fig. 4.

   1975 *Caliceras capillosum* (Barrande); Kolebaba, p. 389, pl. 1, figs 1–4; pl. 2, figs 1–5; pl. 3, figs 1–6; pl. 4, figs 1–3, text-figs 1, 3–4.

   1977 *Caliceras capillosum* (Barrande); Kolebaba, p. 127, text-fig. 1.2.

   1984 '*Caliceras capillosum*' (Barrande); Dzik, pp. 97, 108, 112, text-fig 14.

   1985 *Caliceras capillosum* (Barrande); Kiselev, p. 84, pl. 1, fig. 12.

   1991 *Caliceras capillosum* (Barrande); Kiselev, p. 26, text-figs 2.3–2.4.

Material. 23 fragments, MB.C.10205–28, and 2 apical fragments, MB.C.9591–92, from bed BS, Lochkovian, Filon Douze section.

Diagnosis (after Kolebaba 1975, p. 389). Arionoceratids with angle of expansion of 6–7 degrees; shell surface ornamented with marked fine transverse striae which form shallow sinus on prosiphuncular side; sutures straight, transverse; chambers *c.* 0.3–0.8 of conch cross-section diameter; depth of septal curvature one-quarter of conch cross-section; siphuncle subcentral; diameter of septal perforation *c.* 0.1 of conch cross-section; septal necks very short, suborthochoanitic; connecting ring slightly expanded within chambers; mural and episeptal deposits in apical part of adult specimens.

Description. Conch ornamented with conspicuous, closely spaced, transverse lirae (Pl. 16, figs 5–6) which form shallow sinus on prosiphuncular side, approximately eight in 1 mm in MB.C.10206. Conch cross-section compressed or circular. Of 12 measured specimens, two have circular cross-section. Minimum measured width to height ratio of conch 0.89, average 0.95. Angle of expansion *c.* 7 degrees. No change of angle of expansion during growth visible. Variability of septal spacing relatively low. Septal distance is 0.3–0.4 of conch cross-section diameter. Sutures straight, transverse. In MB.C.10205 chamber length of most adoral chamber is 6.8 mm with conch cross-section of 16.6 mm. Septal perforation 0.9 mm in diameter. Siphuncular segments of entire fragment nearly tubular with cross-section diameter of 1.2 mm. Septal necks very short, loxochoanitic or suborthochoanitic. Length of septal neck of most adoral septum of fragment 0.2 mm. Siphuncular tube slightly constricted at septal perforation. No cameral or endosiphuncular deposits present. Fragments MB.C.9591–9592 are apices with elongate-ovate, slightly bent initial chamber; length of chamber 1.6 and 1.5 mm, and diameter 1.1 and 1.0 mm respectively. Tips of initial chambers acute. Constriction between initial chamber and subsequent conch weakly developed. Conch segment 1.4 mm adoral of initial chamber has angle of expansion of 12 degrees in MB.C.9591.

Remarks. Kolebaba (1975) failed to include information on the shape of the cross-section in the species diagnosis of *A. capillosum.* The specimens figured by Barrande (1874) and Kolebaba (1975) from the type locality and horizon are circular in cross-section. The specimens described here vary in cross-section from circular to compressed and may represent a different species. Because the known apices from bed BS of the Filon Douze section are very similar to that of *A. capillosum* illustrated by Kolebaba (1975), these fragments are referred to this species, which differs from *A. canonicum* mainly in having a transversely ornamented conch.

Occurrence. Ludlow–Lower Devonian; Australia, Czech Republic, Germany, Morocco and Tadzhikstan.

Arionoceras sp. A
Plate 13, figure 30

Material. 7 apical fragments, MB.C.10236–41, from bed OP-M, Ludfordian, Filon Douze section.

Description. Apices with small, elongate-ovate initial chamber of length 0.8 mm, maximum width 0.7 mm. Apical constriction very shallow. Caecum elongate, tubular, crosses entire initial

chamber. Siphuncular position subcentral. Septal perforation *c.* 0.02 mm wide at conch cross-section diameter of 4 mm. Conch adoral of constriction grows at angle of expansion of 12 degrees.

Remarks. The apical fragments described above differ from those of *A. canonicum* in having an initial chamber <1 mm in length.

Genus ADIAGOCERAS gen. nov.

Type species. Adiagoceras taouzense sp. nov., from the Lochkovian of the Filon Douze section.

Derivation of name. A contraction of *Arionoceras* and *Diagoceras. Adiagoceras* is similar to the former with regard to the general conch shape and internally similar to the latter.

Other species included. Orthoceras subnotatum Barrande, 1868 (species combined with *Adiagoceras* below: see 'List of new combinations').

Diagnosis. Short orthocones with angle of expansion of 4–10 degrees; cross-section circular or ovate-compressed; adult cross-section diameter <15 mm; shell ornamented with conspicuous, minute, growth lines that describe shallow lateral lobes; septal distance short, *c.* 0.3 of conch cross-section diameter; sutures oblique, sloping in an adapical direction on prosiphuncular side with shallow lateral lobe; septal necks suborthochoanitic; septal perforations with diameter <0.1 of conch cross-section; siphuncle initially subcentral, central in later growth stages; no endosiphuncular and cameral deposits; initial chamber ovate, sometimes with acute tip; apical first 2 mm slightly cyrtocone with siphuncle on concave side of curvature of growth axis.

Remarks. Adiagoceras is assigned to the Arionoceratidae because of its similarity to *Arionoceras* in general conch shape, the septal necks and shape of the apex. It differs in having a smaller initial chamber, oblique sutures and suborthochoanitic septal necks. It resembles *Clinoceras* Mascke, 1876, in general conch shape, but differs in having suborthochoanitic septal necks and a compressed conch cross-section. The oblique septa and eccentric siphuncle are typical of *Plagiostomoceras* Teichert and Glenister, 1952, which differs in having orthochoanitic septal necks and a spherical apex. The Middle Devonian *Diagoceras* Flower, 1936, is very similar in internal characters, and also displays oblique septa and suborthocoanitic septal necks, but differs in having a large slender conch. The close similarity of *Adiagoceras* to *Diagoceras* sheds new light on the phylogenetic affinities of the latter. Flower (1936, 1939) assigned *Diagoceras* to the Pseudorthoceratidae but remarked (1939, p. 92) that the 'slender siphuncular segments of *Diagoceras* led to its inclusion in the Orthoceratidae.' He considered it to be an early offshoot of *Dolorthoceras*, but it is more likely that it is a late representative of the Arionoceratidae. However, data on the apex are needed in order to solve this problem. *Orthoceras subnotatum* Barrande, 1868 is assigned to *Adiagoceras* herein because of its general similarity to *Adiagoceras taouzense*. Gnoli (1982, pp. 92–93) described the septal necks of *O. subnotatum* as cyrtochoanitic. However, following the definition of Kröger (2006, p. 141), cyrtochoanitic septal necks have brims that are bent more than 180 degrees. The septal necks of *O. subnotatum* are clearly bent <180 degrees and are, therefore, suborthochoanitic.

Occurrence. Lochkovian; Czech Republic, Morocco and Sardinia.

Adiagoceras taouzense sp. nov.
Plate 13, figures 5–7; Plate 14, figures 1–3; Plate 15, figures 1–3

Derivation of name. From Taouz, a military station to the east of the type locality.

Holotype. MB.C.10151 (Pl. 15, figs 1–2).

Type locality and horizon. Filon Douze section, bed OJ-I, Lochkovian.

Other material. 161 fragments of juvenile and mature specimens, MB.C.9958.1–114, 10128.1–21, 10105–6, 10152, and 15 apices, MB.C.10153–10168, from the type locality and horizon; 1 fragment, MB.C.9590, from bed CK, Lochkovian, from the type locality.

Diagnosis. Short orthocones with angle of expansion between 4 and 10 degrees; in earliest growth stages, conch slender; juvenile growth stages have highest angle of expansion; mature growth stages with medium expansion rate of *c.* 7 degrees; adult cross-section diameter *c.* 13 mm; cross-section circular at mature body chamber, slightly compressed in adolescent stages; ornamented with conspicuous, minute, growth lines that describe slight lateral lobes; distance between prominent growth lines in adult specimen *c.* 1 mm; septal distance 0.3 of conch cross-section; sutures oblique, sloping in adapical direction on prosiphuncular side with shallow lateral lobe; septal necks suborthochoanitic; diameter of septal perforation <0.1 of conch cross-section; siphuncle initially subcentral, central in later growth stages; no endosiphuncular and cameral deposits; apical diameter at first septum *c.* 0.4 mm, length of first chamber *c.* 0.5 mm; initial

chamber ovate, sometimes with acute tip; apical first 2 mm slightly cyrtocone with siphuncle on concave side of curvature of growth axis.

Description. Conch nearly straight, most apical 2 mm slightly cyrtocone. Most apical part of conch slender with expansion of <5 degrees. At cross-section diameter of *c.* 2 mm expansion rate increases, resulting in slightly concave conch margins in median section with angle of expansion up to 9 degrees (Text-fig. 22). Holotype with maximum cross-section diameter 8.7 mm, minimum 7.4 mm, has angle of expansion of 6 degrees. Approximately 10 mm adoral of base of mature body chamber (at cross-section diameter of 10 mm in MB.C.10106) expansion rate decreases (compare Text-fig. 23). Mature body chamber of MB.C.10106 has angle of expansion of 7 degrees where conch margin in longitudinal section appears slightly convex. Cross-section slightly compressed or circular. Conch surface ornamented by faint growth lines of varying relief (Pl. 14, figs 1, 3). Distance between prominent growth lines in MB.C.10105 1 mm. Growth lines describe shallow lateral lobe. Septal spacing in MB.C.10106 with cross-section diameter of 9 mm, 1.2 mm; in holotype with cross-section diameter of 7.4 mm, 1.2 mm. Mature septal crowding occurs in MB.C.10106 at cross-section diameter of 9.8 mm with septal distance of most adoral septa 0.8 mm, where septal perforation is central and 1 mm in diameter. Septal necks suborthochoanitic (Pl. 15, figs 1–3). In holotype at cross-section diameter of 7.4 mm centre of septal perforation is 3.2 mm from conch margin and 0.4 mm in diameter. Apical diameter at first septum *c.* 0.4 mm; length of first chamber *c.* 0.5 mm. Shape of initial chamber variably ovate, sometimes with acute tip, sometimes rounded. Apical first 2 mm slightly cyrtocone with siphuncle on concave side of curvature of growth axis (Pl. 13, figs 5–7). In MB.C.10155 second to fourth chambers closely spaced, only 0.1 mm apart. Height of fifth chamber 0.17 mm; of sixth, 0.2 mm. Septal perforation at most apical chambers half-way between centre and margin of conch. Position of septal perforation shifts in later septa increasingly towards centre. Most apical septum of apex oblique with suborthochoanitic septal neck.

Remarks. Although the angle of expansion can vary considerably between specimens (Text-fig. 22), it is impossible to distinguish two or more species of *Adiagoceras* using this character. *Adiagoceras taouzense* is the predominant cephalopod in bed OJ-I of the Filon Douze section, which is composed almost entirely of fragments of all sizes of this species. The apex of *A. taouzense* differs from the apices described and illustrated by Serpagli and Gnoli (1977, text-fig. 5), which they assigned to *Protobactrites*? sp., in having a larger apical expansion rate and oblique septa. Moreover, these Silurian apices have a more pronounced, constricted, and more elongate initial chamber. The initial chambers of species of *Arionoceras*, except *A. submoniliforme* (Meneghini, 1857), have different dimensions and differ in shape, mainly in having an acute tip; they are also clearly constricted in later growth stages. No clear constriction occurs in *Adiagoceras*. *Orthoceras subnotatum* Barrande, 1868 differs from *A. taouzense* in having longer septal necks and an ovate-compressed cross-section.

Genus PARAKIONOCERAS Foerste, 1928

Type species. *Orthoceras originale* Barrande, 1868, from the mid Silurian of Bohemia, Czech Republic.

Diagnosis (after Sweet 1964*b*, p. K229). Slightly cyrtoconic conch which increases moderately; conch surface vertically incised at regular intervals by very narrow, shallow grooves; cross-section circular; septa transverse and widely spaced; septal necks short and loxochoanitic; siphuncle thin and subcentrally positioned on convex side of curvature of conch growth axis; siphuncular segments slightly constricted at septal necks, otherwise essentially cylindrical; no siphuncular deposits; presence or absence of cameral deposits not known.

Occurrence. Wenlock–Lochkovian; world-wide.

Parakionoceras originale (Barrande, 1868)
Plate 14, figures 4–5

1855 *Orthoceras originale* Barrande, p. 450 [nomen nudum].
* 1868 *Orthoceras originale* Barrande, pl. 267, figs 1–20.
1874 *Orthoceras originale* Barrande, pp. 206–209.
1882 *Orthoceras originale* Barrande; Blake, p. 110, pl. 7, figs 5, 5a, 10.
1914 *Orthoceras originale* Barrande; Vidal, p. 9, pl. 3, fig. 8.
1928 *Parakionoceras originale* (Barrande); Foerste, p. 314.
1972 *Parakionoceras originale* (Barrande); Barskov, p. 47, pl. 1, figs 9–10; pl. 2, fig. 14.
1977 *Parakionoceras originale* (Barrande); Kolebaba, p. 133, pl. 2, fig. 5, text-fig. 5.
1977 *Parakionoceras originale* (Barrande); Serpagli and Gnoli, p. 180, pl. 7, fig. 5.
1978 *Parakionoceras originale* (Barrande); Zhuravleva, p. 67, pl. 1, figs 9–10.
non 1984 *Parakionoceras originale* (Barrande); Dzik, pp. 108, 122, pl. 29, figs 1–2; pl. 30, figs 3–4, text-fig. 42.17 [*fide* Turek and Marek 1986].
1986 *Parakionoceras originale* (Barrande); Turek and Marek, p. 252.

EXPLANATION OF PLATE 13

Apices from the Filon Douze section.
Fig. 1. *Sphaerorthoceras* sp. B, MB.C.9511; bed BS, Lochkovian.
Fig. 2. *Merocycloceras* sp. A, MB.C.10179; bed OJ-I, Lochkovian.
Fig. 3. *Merocycloceras* sp. B, MB.C.9787; bed KMO-I, Pragian.
Fig. 4. *Parasphaerorthoceras* sp. H Ristedt, 1968; MB.C.9521; bed BS, Lochkovian.
Figs 5–7. *Adiagoceras taouzense* sp. nov.; bed OJ-I, Lochkovian. 5, MB.C.10136. 6, MB.C.10155. 7, MB.10153.
Fig. 8. *Bactrites*? sp., MB.C.9764; bed KMO-III, Pragian.
Figs 9–10. *Devonobactrites obliqueseptatus* (Sandberger and Sandberger, 1852); bed EF, Zlíchovian, Emsian. 9, MB.C.9548. 10, MB.C.9549.
Figs 11–12. *Kopaninoceras*? sp.; bed OP-M, Ludfordian, Ludlow. 11, MB.C.10080. 12, MB.C.10079.
Fig. 13, 15. *Akrosphaerorthoceras* sp.; bed BS, Lochkovian. 13, MB.C.9627. 15, MB.C.9596.
Fig. 14. *Merocycloceras*? sp. C, MB.C.9794; bed BS, Lochkovian.
Figs 16–17. *Michelinoceras* sp.; bed OP-M, Ludfordian, Ludlow. 16, MB.C.10078. 17, MB.C.10077.
Figs 18–19. *Hemicosmorthoceras aichae* sp. nov.; MB.C.9586, holotype; bed OJ-I, Lochkovian. 18, lateral view. 19, view on convex side of growth axis.
Fig. 20. *Sphaerorthoceras* sp. A, MB.C.9510; bed BS, Lochkovian.
Fig. 21. *Parasphaerorthoceras* sp. B Ristedt, 1968; MB.C.9524; bed BS, Lochkovian.
Fig. 22. *Parasphaerorthoceras* sp. G Ristedt, 1968; MB.C.9517; bed PK, Lochkovian.
Figs 23–29. *Arionoceras canonicum* (Meneghini, 1857); note the variability in length of apical chambers. 23–28, bed OP-M, Ludfordian, Ludlow. 23–24, MB.C.10234; mould of apical chambers showing negative impression of empty spaces of siphuncle and chambers without cameral deposits. 23, lateral view. 24, prosiphuncular view. 25–26, MB.C.10235; mould of apical chambers showing shape of long ceacum. 25, lateral view. 26, prosiphuncular view. 27, MB.C. 10231; lateral view. 28, prosiphuncular view. 29, MB.C.9591, bed BS, Lochkovian; lateral view.
Fig. 30. *Arionoceras* sp. A, MB.C.10240; bed OP-M, Ludfordian, Ludlow; mould of apical chambers showing shape of caecum, lateral view.
All × 15.

PLATE 13

KRÖGER, nautiloids

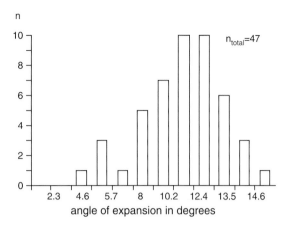

TEXT-FIG. 22. Frequency distribution of angle of expansion of 47 specimens of *Adiagoceras taouzense* gen. et sp. nov. Note the single frequency peak at about 12 degrees.

1988 *Parakionoceras originale* (Barrande); Gómez-Alba: p. 352, pl. 173, fig. 5.

Material. 4 fragments, MB.C.9518-19, 9948, 10132, from bed OSP-II, Přídolí; 1 fragment, MB.C.9520, from bed BS; 1 fragment, MB.C.9957, from bed CK; 2 specimens, MB.C.9516–17, from bed OJ-I, Lochkovian, Filon Douze section.

Diagnosis (after Barskov 1972, p. 47). *Parakionoceras* with angle of expansion between 10 and 14 degrees, sometimes greater; ornamented with *c.* 60 longitudinal lacunae around circumference; three septa at distance similar to conch cross-section diameter; diameter of siphuncle <0.1 of conch cross-section; siphuncular segments nearly tubular, slightly constricted at septal perforation.

Description. Largest fragment has cross-section diameter of 21 mm (MB.C.10132), representing nearly tubular part of body chamber. Distance between individual lacunae *c.* 1.1 mm. In MB.C.9518, diameter cannot be determined; however, distance between individual lacunae 1.4 mm, with reference to original fragment diameter of *c.* 26 mm. Fragment of phragmocone MB.C.9957 16 mm long, cross-section diameter 7–9 mm (angle of expansion 11 degrees), with slightly compressed cross-section. Chamber distance 3.1 mm. Position of siphuncle subcentral with diameter of 0.9 mm.

Occurrence. Wenlock–Lochkovian; Czech Republic, Kazakhstan, Morocco, Spain and UK.

Family DAWSONOCERATIDAE Flower, 1962

Genus ANASPYROCERAS Shimizu and Obata, 1935*a*

Type species. Orthoceras anellus Conrad, 1843, from the Beloit Member, Black River Formation, Ordovician of Mineral Point, Wisconsin, USA.

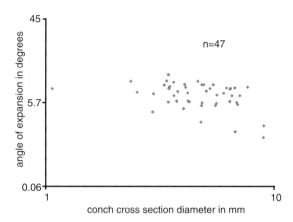

TEXT-FIG. 23. Angle of expansion and average cross-section diameter of 47 fragments of *Adiagoceras taouzense* gen. et sp. nov. Note the tendency towards lower angles of expansion in the specimens with a larger cross section. Axes scaled logarithmically.

Diagnosis (after Flower 1946, p. 139; Sweet 1964*b*, p. K230). Dawsonoceratidae with slender orthoconic conchs; surface ornamented with conspicuous transverse undulations, longitudinal lirae, and faint transverse lirae; cross-section circular; septa transverse and closely spaced; septal necks suborthochoanitic; siphuncle subcentral; siphuncular segments tubular or slightly expanded at septal necks; siphuncular and endocameral deposits not known.

Remarks. For discussion of the family assignment of *Anaspyroceras* within the Dawsonoceratidae, see Kröger and Isakar (2006, p. 153).

Occurrence. Middle Ordovician–Early Devonian; Czech Republic, Morocco and North America.

Anaspyroceras sp.
Plate 7, figure 11; Plate 10, figure 10; Plate 15, figure 9

Material. 1 specimen, MB.C.9680, from bed KMO-III, Pragian, Filon Douze section.

Description. Fragment slightly cyrtoconic, 27 mm long, circular cross-section with diameter 9–11 mm (angle of expansion 4 degrees). Conch ornamented with undulations at distance of 4 mm, longitudinal lirae 2.5 mm apart (nine occur across circumference), and closely spaced transverse lirae. Sutures transverse. Distance between septa 3.5 mm. Septal concavity shallow. Septal perforations subcentral, diameter 1.1 mm at adoral end of fragment (0.1 of conch-cross-section diameter). Septal necks suborthocoanitic. Siphuncular tube sightly expanded within chambers and constricted at septal foramen.

Remarks. This fragment is assigned to *Anaspyroceras* because it displays suborthochoanitic septal necks and a siphuncle that is only slightly expanded within the chambers. It differs from *A. pseudocalamiteum* (Barrande, 1851) in having a larger angle of expansion, a slightly cyrtoconic conch, and a longitudinal ornamentation that consists only of equally raised lirae.

Indet. Dawsonoceratidae
Plate 15, figure 4

Material. 1 specimen, MB.C.10083, from bed KMO-I, Pragian, Filon Douze section.

Description. Fragment of slightly cyrtoconic phragmocone 13 mm long; conch cross-section nearly circular with diameter 5.9–11.6 mm (angle of expansion of 24 degrees); 12 chambers present. Septal spacing close; minimum 1.2 mm (at adapical end of fragment), maximum 1.8 mm (at cross-section diameter of 13 mm). Siphuncle subcentral on convex side of growth axis. Diameter of septal perforation 0.5 mm at cross-section diameter of 10 mm where diameter of siphuncle is 1.1 mm. Septal necks achoanitic. Siphuncular segments essentially tubular with narrow constrictions at septal perforation. At adoral and adapical surfaces of septa, large area of adnation of wide siphuncle. No endosiphuncular and cameral deposits.

Remarks. The internal structures of this fragment are very distinctive. The septal necks are achoanitic and the nearly tubular siphuncle is strongly constricted at the septal necks forming large areas of adnation. The very short septal necks and the tubular siphuncle are diagnostic of the Dawsonoceratidae. However, no generic or specific determination is possible on the basis of this single apical fragment because it does not show any surface characters.

Family GEISONOCERATIDAE Zhuravleva, 1959

Genus ANGEISONOCERAS gen. nov.

Type species. *Orthoceras davidsoni* Barrande, 1870, from the Upper Silurian of Lochkov, Bohemia, Czech Republic.

Derivation of name. Contraction of the Latin *angustus*, narrow, with reference to the closely spaced transverse sculpture, and *Geisonoceras* Hyatt, 1884, which this genus otherwise resembles.

Other species included. *Angeisonoceras reteornatum* sp. nov.; *Orthoceras consocium* Barrande, 1870; *O. festians* Barrande, 1870; *O. imminutum* Barrande, 1870; *O. inchoatum* Barrande, 1868; *O. libens* Barrande, 1870; *O. limatum* Barrande, 1870; *O. squamatulum* Barrande, 1868 (all *Orthoceras* species combined with *Angeisonoceras* below: see 'List of new combinations').

Diagnosis. Slowly widening orthocones or cyrtocones circular or subcircular in cross-section with straight transverse or slightly oblique sutures; surface ornamented with narrow transverse ridges and valleys approximately the same width, in some cases with longitudinal striae; aperture transverse with shallow hyponomic sinus; siphuncle central; septal necks suborthocoanitic; siphuncular segments expanding slightly within camerae; small, adorally attenuated, annulosiphonate deposits in adapical siphuncular segments; episeptal and hyposeptal deposits present.

Remarks. This genus comprises species that are similar to *Geisonoceras* Hyatt, 1884, in general conch shape and ornamentation. However, in that genus, the surface has broadly transverse bands with fine growth lines; *Angeisonoceras* differs in having closely spaced transverse ridges. Although the sculpture of *Geisonoceras rivale* Barrande, 1859, is very conspicious and the generic diagnosis of Sweet (1964*b*) refers to this ornamentation, the genus is widely used as receptacle for Geisonoceratidae with a generally transverse ornamentation. Of the 50 or more species assigned to *Geisonoceras* when it was erected, several should probably be assigned to *Angeisonoceras*, but of these, I list only those illustrated by Barrande (1860–70).

Occurrence. Early Devonian; Czech Republic, Morocco and Sardinia.

Angeisonoceras reteornatum sp. nov.
Plate 7, figure 15; Plate 10, figure 15; Plate 15, figure 7

Derivation of the name. Latin, *rete*, net, with reference to the characteristic ornamentation of the species.

Holotype. MB.C.9946 (Pl. 10, fig. 15; Pl. 15, fig. 7).

Type locality and horizon. Filon Douze section, bed KMO-II, Pragian.

Other material. 4 specimens, MB.C.9582–83 and 10092–94, from the type locality and horizon.

Diagnosis. Orthocones with slightly compressed cross-section; angle of expansion 5–8 degrees; ornamented with fine, closely spaced, transverse ridges; ridges faintly undulating, crossed by irregularly spaced, longitudinal, smooth striae; sutures straight, transverse at distance of *c.* 0.3 of conch cross-section; septal perforations subcentral with diameter of 0.1 of conch cross-section; septal necks suborthochoanitic.

Description. Holotype is fragment of body chamber and three chambers of phragmocone, 34 mm long; cross-section slightly compressed, 21–26 mm in diameter (angle of expansion 8 degrees). Conch surface only partially preserved with transverse ridges. Seven irregularly elevated ridges occur at distance of

1 mm, some more elevated than others, faintly irregularly festooned. Irregularly spaced, longitudinal, smooth striae cross
transverse ornamentation. Septa transverse. Height of three
chambers measured from adapical to adoral, 7.3, 7.7 and
7.5 mm. Septal perforation subcentral with diameter 2.5 mm at
adapical septum. Septal necks suborthochoanitic, 1.3 mm long.

MB.C.10092, fragment of phragmocone 26 mm long; adoral
diameter 15.3–17.8 mm (angle of expansion 5 degrees), ornamented with faintly festooned transverse ridges (Pl. 7, fig. 15),
eight occurring at distance of 1 mm, and longitudinal irregularly
spaced smooth striae, crossing transverse ornamentation, forming faint rectangular pattern. Siphuncle subcentral. Diameter of
septal perforation at most adapical septum 1.9 mm.

Remarks. This species differs from all other species of the
genus in having transverse striations that are crossed by
longitudinal smooth striae.

Genus TEMPEROCERAS Barskov, 1960

Type species. Orthoceras ludense Sowerby, 1839, from the Gorstian (early Ludlow) of Ludlow, Shropshire, England; by subsequent designation of Holland (2000).

Diagnosis (after Zhuravleva 1978, p. 75). Smooth orthocones with circular cross-section; septal spacing moderate;
sutures straight or very slightly lobate; siphuncle subcentral and with segments slightly expanded within
chambers; during ontogeny siphuncle position shifts successively from centre and relative siphuncular diameter
decreases; septal necks suborthocoanitic–orthochoanitic;
endosiphuncular deposits form annuli at septal necks;
cameral deposits on prosiphincular side of conch.

Occurrence. Late Silurian–Early Devonian; world-wide.

Temperoceras? ludense (Sowerby, *in* Murchison 1839)
Plate 12, figure 7

1839 *Orthoceras ludense* Sowerby, p. 619, pl. 9, fig. 1a–b.
1867–70 *Orthoceras temperans* Barrande, pl. 230, figs 7–9;
pl. 327, figs 1–2; pl. 382; pl. 451, figs 1–3.
1868 *Orthoceras omnium* Barrande, pl. 215, figs 2–3.
1870 *Orthoceras dahlii* Barrande, pl. 440, figs 7–10.
1874 *Orthoceras temperans* Barrande, p. 658.
1876 *Orthoceras ludense* Sowerby; Remelé, p. 425.
1882 *Orthoceras ludense* Sowerby; Blake, p. 156, pl. 10,
figs 1, 3–5, 7.
1925 *Orthoceras temperans* Barrande; Heller, p. 243.
1962 *Temperoceras temperans* (Barrande); Balashov and
Zhuravleva, pl. 12, fig. 2.
1966 *Temperoceras temperans* (Barrande); Babin, p. 324,
pl. 13, fig. 5; pl. 14, fig. 1a–c.
1972 *Temperoceras temperans* (Barrande); Barskov, p. 49,
pl. 1, figs 1–2.
1999 *Temperoceras temperans* (Barrande); Histon and
Gnoli, pp. 384, 388, tables 1–2.
2000 *Temperoceras ludense* (Sowerby); Holland, p. 120,
pl. 1, figs 3–4, 6–7, text-fig. 1 [with synonymy].

Material. 6 specimens, MB.C.9512, 9508, 9576.1–3, 9799, from
bed OP, Ludfordian; 6 specimens, MB.C.9571.1–2, 9577.1–4,
from bed OJ-I, Lochkovian, Filon Douze section.

Diagnosis. Conch straight, slender, with angle of expansion of *c.* 7 degrees; conch cross-section circular; conch
surface smooth or very shallowly transversely undulating
at body chamber of late growth stages; irregularly spaced

EXPLANATION OF PLATE 14

Figs 1–3. *Adiagoceras taouzense* sp. nov., bed OJ-I, Lochkovian. 1, 3, MB.C.10105. 1, entire specimen, lateral view. 3, detail of conch
surface showing conspicuous growth lines; scale bar represents 1 mm. 2, MB.C.10157; adapical view.

Figs 4–5. *Parakionoceras originale* (Barrande, 1868); MB.C.9957; bed CK, Lochkovian; details of the conch surface with characteristic
longitudinal striae; scale bar represents 1 mm.

Figs 6–9. *Devonobactrites obliqueseptatus* (Sandberger and Sandberger, 1852); bed EF, Zlíchovian, Emsian. 6, MB.C9545; body chamber
with large, healed, shell breakage at peristome; ventral view. 7, MB.C9546; section of phragmocone showing the closely spaced,
oblique septation. 8–9, MB.C9547; body chamber. 8, lateral view. 9, adapical view.

Figs 10–11. *Lobobactrites ellipticus* (Frech, 1897); MB.C.9580; bed F, *P. partitus* conodont Biozone, early Eifelian. 10, lateral view; 11,
adoral view.

Fig. 12. *Bactrites gracile* (Blumenbach, 1803), MB.C.9945; bed F, *P. partitus* conodont Biozone, early Eifelian; natural section of ventral
part of conch.

Fig. 13. *Kionoceras? doricum* (Barrande 1868); MB.C.10101; bed OP-M, Ludfordian, Ludlow; lateral view.

Fig. 14. *Temperoceras aequinudum* sp. nov; MB.C.9505, holotype; bed KMO-III, Pragian; median section.

Fig. 15. *Devonobactrites* sp.; MB.C.1095; bed EF, Zlíchovian, Emsian; lateral view.

Fig. 16. *Sichuanoceras zizense* sp. nov.; MB.C.9509, holotype; bed OJ-I, Lochkovian; median section; note the conspicuous annular
endosiphuncular deposits.

All specimens from the Filon Douze section and × 2 except where indicated by scale bar.

PLATE 14

KRÖGER, nautiloids

conspicuous growth lines; septal distance 0.2–0.4 of conch cross-section; sutures straight, transverse; depth of septal curvature equals depth of chambers; septal perforation central or subcentral with diameter 0.1–0.2 of conch cross-section; siphuncular segments slightly expanded within chambers; septal necks suborthochoanitic, length about half diameter of septal perforation; endosiphuncular annuli at septal perforation; episeptal, mural, and hyposeptal deposits.

Description. MB.C.9799 from bed OP is fragment of two chambers 30 mm in total length; conch cross-section circular, diameter 45–47 mm. Depth of septal curvature 14 mm. Septal perforation close to conch centre. Centre of septal perforation of adapical septum 22 mm from conch margin. Diameter of septal perforation 8 mm. Septal necks suborthochoanitic, 3 mm long. Siphuncular segments slightly expanded, maximum diameter at midlength of 10 mm. At septal perforation endosiphuncular annuli present.

MB.C.9508, from bed OJ-I is fragment of two chambers 31 mm in total length; conch cross-section circular, diameter 47–49 mm. Depth of septal curvature 16 mm. Septal perforation close to conch centre. Centre of septal perforation of adapical septum 24 mm from conch margin. Diameter of septal perforation 9 mm. Septal necks suborthochoanitic, 1.9 mm long. Siphuncular segments slightly expanded, maximum diameter at midlength of 9.4 mm. At septal perforation endosiphuncular annuli present.

MB.C.9512, from bed OJ-I is fragment of two chambers 29 mm in total length; conch cross-section circular, diameter 36–38 mm. Depth of septal curvature 11 mm. Septal perforation close to conch centre. Centre of septal perforation of adapical septum 17 mm from conch margin. Diameter of septal perforation 5.6 mm. Septal necks suborthochoanitic, 1.3 mm long. Connecting ring not preserved. No cameral and endosiphuncular deposits present.

Remarks. The specific diagnosis is based on Holland (2000, pp. 120–121) and the illustrations of *Orthoceras temperans* in Barrande (1867–70). Barrande's material shows septal perforation of comparatively varied diameter, ranging between 0.13 and 0.19 mm, as opposed to 0.13–0.14 mm in Holland's specimens. Thus, there is the possibility that at least some of the specimens figured by Barrande may be a separate species and that *O. temperans* may encompass specimens with a wider siphuncle. However, more data are needed to confirm this conclusion. I follow Holland (2000) herein and regard *O. temperans* and *O. ludense* as synonyms. *Temperoceras ludense* differs from all other species of the genus in having a comparatively central siphuncle combined with moderate septal spacing and a deep septal curvature.

Occurrence. Wenlock (Homerian)–Lochkovian; Czech Republic, England, France, erratics in northern Germany, Kazakhstan, Morocco and Sardinia.

Temperoceras aequinudum sp. nov.
Plate 14, figure 14

2008 *Temperoceras* sp.; Klug *et al.*; pp. 45–46, pl. 6, figs 19–22.

Derivation of name. Latin, *nudus*, naked, and with reference to *Orthoceras nudum* Barrande, 1870, which is similar in septal spacing, septal neck shape, and position of siphuncle.

Holotype. MB.C.9505 (Pl. 14, fig. 14).

Type locality and horizon. Filon Douze section, bed KMO-III, Pragian.

Other material. 2 specimens, MB.C.9575.1–2, from bed OJ-II; 10 specimens, MB.C.9755, 9758.1–9, from bed KMO-I; 27 specimens, MB.C.5759.1–27, from bed KMO-II; 8 specimens, MB.C.5763.1–8, from bed KMO-III; 38 specimens, MB.C.5767.1–38, from bed EF, Lochkovian–Pragian, Filon Douze section.

Diagnosis. Conch straight with comparatively high angle of expansion of more than 5 degrees; conch cross-section circular; septal distance *c.* 0.3 of conch cross-section; sutures straight, transverse; depth of septal curvature *c.* 0.5–0.7 (or less) of depth of chambers; septal perforation subcentral with diameter 0.10–0.15 of conch cross-section; siphuncular segments strongly expanded within chambers; septal necks suborthochoanitic, less than half diameter of septal perforation in length; endosiphuncular annuli at septal perforation.

Description. MB.C.9506 is largest fragment of phragmocone with 11 complete chambers and parts of shell surface, which is faintly irregularly undulating, otherwise smooth. Conch margin slightly convex in longitudinal section; cross-section circular. Fragment 115 mm long, cross-section diameter 32–53 mm (angle of expansion 10 degrees). Septal distance at cross-section diameter of 33 mm is 10 mm (0.3 of conch cross-section) where depth of septal curvature is *c.* 7 mm. Septal perforation subcentral 15 mm from conch margin. Diameter of septal perforation 3.3 mm at conch cross-section diameter of 30 mm. Siphuncular segments strongly expanded within chambers with siphuncular diameter at mid-length between septa *c.* 6.5 mm. Shallow episeptal and mural deposits occur on prosiphuncular side.

Second largest specimen, MB.C.9956, is fragment of phragmocone with four complete chambers; total length 64 mm. Conch margin slightly convex in longitudinal section; cross-section circular, 43–49 mm in diameter (angle of expansion 5 degrees), chamber distance varies between 0.3 (most adoral chamber) and 0.23 (most adapical chamber) of conch cross-section. Depth of septal curvature 0.75–0.8 of chamber distance. Septal perforation subcentral with centre of perforation of most adoral septum 23 mm from conch margin. Diameter of septal perforation of most adoral septum 5.8 mm (0.12

of conch cross-section). Septal necks suborthochoanitic, 1 mm long at most adoral septum.

MB.C.9505 is fragment of phragmocone with seven complete chambers; total length 56 mm; cross-section diameter 25–32 mm (angle of expansion 7 degrees). Septal perforation subcentral with diameter of 4.3 mm at most adoral septum (0.13 of conch cross-section). Endosiphuncular annuli present at septal perforation.

Remarks. Temperoceras aequinudum is distinguished from other species of the genus in having a combination of comparatively narrow septal spacing and a nearly central siphuncle that is strongly expanded within chambers. *T. nudum* Barrande (1870) differs in having a wider siphuncular diameter of 0.2 of conch cross-section and a lower angle of expansion of 5 degrees.

Temperoceras migrans (Barrande, 1868)
Plate 15, figures 6, 8

   non 1868 *Orthoceras migrans* Barrande, pl. 212, figs 1–4.
  * 1868–70 *Orthoceras migrans* Barrande, pl. 222, figs 1–2;
             pl. 309, figs 1–4; pls 348, 377.
     1874 *Orthoceras migrans* Barrande, p. 643.
     1878 *Orthoceras* cf. *migrans* Barrande; Kayser, p. 71,
             pl. 10, fig. 6.
     1968 *Michelinoceras migrans* (Barrande); Balashov and
             Kiselev, p. 8, pl. 1, fig. 1.
     1972 *Temperoceras migrans* (Barrande); Barskov, p. 49,
             pl. 4, fig. 1.
     1998 *Orthoceras migrans* (Barrande); Histon, p. 108.

Material. 24 specimens, MB.C.9759.1–22, 9775, 10099, from bed KMO-I; 2 specimens, MB.C.5760–61, from bed KMO-II; 3 specimens, MB.C.9507, 9950–52, from bed KMO-III; 1 specimen, MB.C.9592, from bed KMO-IV; 3 specimens, MB.C.9769.1–3, from bed KMO-V; Pragian–Dalejian, Filon Douze section.

Diagnosis. Conch straight, slender, angle of expansion 2–3 degrees; cross-section circular; surface smooth; septal distance 0.5–1 of conch-cross-section; sutures straight, transverse; depth of septal curvature *c.* 0.5–0.7 of depth of chambers; septal perforation subcentral or slightly eccentric with diameter 0.10–0.16 of conch cross-section; siphuncular segments nearly tubular, very slightly expanded within chambers; septal necks suborthochoanitic, less than half of diameter of septal perforation in length; endosiphuncular annuli at septal perforation.

Description. MB.C.9950 has smooth shell, circular cross-section and very low angle of expansion; fragment 41 mm long, cross-section diameter 17.5–19.5 mm (angle of expansion 2.8 degrees); five complete chambers (septal distance *c.* 0.5 of conch cross-section). Sutures straight, transverse. Depth of curvature of most apical septum 5 mm. Septal perforation subcentral. Centre of septal perforation of most apical septum 8.6 mm from conch

margin where diameter of septal perforation is 2.5 mm (0.14 of conch cross-section). Septal necks suborthochoanitic, 0.7 mm long. Siphuncular segments are nearly tubular. Thin hyposeptal deposits present on prosiphuncular side.

MB.C.10099 (Pl. 15, fig. 6) displays three complete chambers; total length 23 mm; conch cross-section circular, diameter 14.9–15.1 mm. Depth of curvature of most apical septum 3.6 mm. Septal perforation subcentral. Centre of septal perforation of most apical septum 6.8 mm from conch margin where diameter of septal perforation is 1.7 mm (0.12 of conch cross-section). Septal necks suborthochoanitic, 0.4 mm long. Siphuncular segments are nearly tubular. Thin endosiphuncular annuli at septal perforation.

Specimens MB.C.9769.1–3 differ from Pragian specimens in having a slightly wider siphuncle, *c.* 0.16 of conch cross-section, and a slightly closer septal spacing (distance between septa 0.4 of conch cross-section).

Remarks. The diagnosis above is based on the illustrations of Barrande (1867–70). *Temperoceras migrans* is distinguished from other species of the genus in its combination of wide septal spacing, narrow, nearly tubular siphuncle and low angle of expansion. The four Emsian specimens have a slightly wider siphuncle and slightly closer septal spacing than the Pragian specimens. However, given the variation of the Bohemian specimens illustrated by Barrande, this variability is regarded as intraspecific.

Occurrence. Ludlow–Lochkovian; Carnic Alps, Austria, Czech Republic, Harz Mountains (Germany), Kazakhstan, Morocco, Sardinia and Ukraine.

Family KIONOCERATIDAE Hyatt, 1900

Genus KIONOCERAS Hyatt, 1884

Type species. Orthoceras doricum Barrande, 1868, from the Upper Silurian of Bohemia, Czech Republic.

Diagnosis (after Sweet 1964*b*, p. K229). Conch slender to rapidly expanding, apically curved and longitudinally fluted with circular to subcircular cross-section; ornament consists of prominent longitudinal ribs separated by concave interspaces and, in most species, less conspicuous longitudinal and transverse lirae and striae; early formed part of test may be faintly undulating; siphuncle central to subcentral; septal necks short, suborthochoanitic; siphuncular segments tubular, weakly expanded within chambers; no endosiphuncular or cameral deposits.

Occurrence. The time range and geographic distribution of the genus is difficult to assess at present because it serves as a 'wastebasket' taxon for longitudinally ornamented orthocones with a variety of internal characters. It seems to be restricted to the Silurian.

Kionoceras? *doricum* (Barrande, 1868)
Plate 14, figure 13

 * 1868   *Orthoceras doricum* Barrande; pl. 269, figs 1–32.
   1928   *Orthoceras doricum* Barrande; Foerste, p. 285.
   1964   *Kionoceras doricum* (Barrande); Sweet, p. K229,
          fig. 159.1a–c.
   1977   *Kionoceras doricum* (Barrande); Serpagli and Gnoli,
          p. 80, pl. 7, fig. 6a–c.
   1979   *Kionoceras doricum* (Barrande); Gnoli *et al.*, p. 418,
          table 4.
non 1984   *Kionoceras doricum* (Barrande); Dzik, pp. 121, 126,
          pl. 36, figs 3, 5, text-figs 48, 49.19 [*fide* Turek and
          Marek, 1986].
   1999   *Kionoceras doricum* Histon and Gnoli, pp. 384, 388,
          tables 1–2.

Material. 4 specimens, MB.C.100101–104, from nodules in bed OP-M, Ludfordian, Filon Douze section.

Diagnosis. Conch slightly curved in early growth stages, straight and slender in late growth stages; angle of expansion 5–9 degrees; adult size *c.* 30 mm in cross-section; ornamented with raised longitudinal lirae and fine transverse striae; between first order longitudinal lirae up to five second order lirae, 20–25 first order lirae around circumference; transverse striae very fine and festooned; in early growth stages conch slightly undulating; septal distance *c.* 0.3 of conch cross-section; sutures straight, transverse; siphuncle central or subcentral with diameter 0.4 of conch cross-section; siphuncular segments tubular, slightly expanded within chambers; septal necks short, suborthochoanitic.

Description. All four specimens only show surface characters. MB.C.10101 is 19 mm in diameter and has 15 first order lirae around circumference, which are distinctively raised above an otherwise nearly smooth surface. One or two second order lirae occur between two first order lirae. Conch straight, without undulation.

MB.C.10102 is 33 mm long; cross-section diameter 13–18 mm (angle of expansion 9 degrees). About 16 first order raised lirae occur around circumference and about six second order lirae occur between two first order lirae. At adoral end of fragment, a third order of longitudinal lirae occurs, which are very closely spaced, *c.* 8 per 1 mm.

Remarks. The generic diagnosis is based on Barrande's illustrations. The specimens here are very similar to those illustrated by Barrande (1868) because the longitudinal striation consists of first and second order striae. However, the distance between the first order striae of the Filon Douze specimens is greater than that of Barrande's. A wide range of ornament occurs in the type material but it is unclear if this reflects an intraspecific variability. Thus, a satisfactory species designation of the Filon Douze material will only be possible after revision of the type material.

Occurrence. Ludlow; Czech Republic, Sardinia and Morocco.

Family SICHUANOCERATIDAE Zhuravleva, 1978

Remarks. Zhuravleva (1978, p. 168) assigned the Sichuanoceratidae to the Bajkaloceratida. However, neither the Silurian–Devonian Sichuanoceratidae nor the Offleioceratidae have anything in common with the Early Ordovician Bajkaloceratidae except for the strong endosiphuncular annuli that form rings with a subquadratic or rectangular shape in median section. In *Baikaloceras* Balashov, 1962, the connecting rings are concave, the endosiphuncular deposits have longitudinal lamellae, and the chambers are short. These characters are typical of Early Ordovician ellesmerocerids, endocerids and intejocerids.

EXPLANATION OF PLATE 15

Median sections of Orthocerida and Pseudorthocerida from the Filon Douze section.

Figs 1–3. *Adiagoceras taouzense* sp. nov.; bed OJ-I, Lochkovian. 1–2, MB.C.10151, holotype. 1, section of three most adoral chambers and base of body chamber. 2, detail of septal perforation and very short suborthochoanitic septal necks. 3, MB.C.10106; detail of septal perforation and septal necks in chambers of juvenile specimen; note angular suborthochoanitic septal necks.

Fig. 4. Indet. Dawsonoceratidae; MB.C.10083; bed KMO-I, Pragian; note the short septal necks and the wide area of adnation.

Fig. 5. *Kopaninoceras*? *jucundum* (Barrande, 1870); MB.C.10085; bed OJ-I, Lochkovian; section shows shape of siphuncular segment and septal neck.

Figs 6, 8. *Temperoceras migrans* (Barrande, 1868). 6, MB.C.10099; bed KMO-I, Pragian; details of shape of siphuncular segment and septal necks. 8, MB.C. 9950; bed KMO-III, Pragian; section shows position and shape of siphuncle and suborthochoanitic septal necks; scale bar represents 10 mm.

Fig. 7. *Angeisonoceras reteornatum* gen. et sp. nov.; MB.C.9946, holotype; bed KMO-II, Pragian; note the orthochoanitic septal necks; scale bar represents 10 mm.

Fig. 9. *Anaspyroceras* sp.; MB.C.9680; bed KMO-III, Pragian; detail of siphuncular segment and septal necks.

Scale bars represent 1 mm except where indicated.

PLATE 15

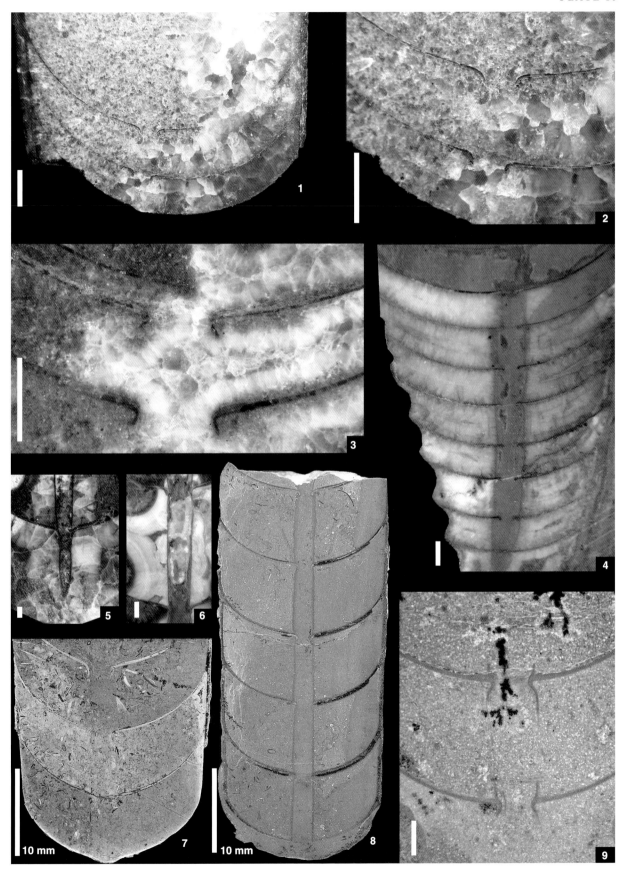

KRÖGER, Orthocerida, Pseudorthocerida

In contrast, the Sichuanoceratidae have relatively long chambers, tubular or expanded siphuncular segments, and homogenous endosiphuncular annuli. Tsou (1966) placed *Sichuanoceras* Chang, 1962, in the Pseudorthoceratidae. This cannot be accepted because the endosiphuncular deposits in pseudorthoceratids are typically arranged strongly asymmetrically in a longitudinal direction along the septal perforation. In *Sichuanoceras* and its allies the endosiphuncular deposits consist of annuli that are positioned nearly symmetrically around the septal perforation as in geisonoceratids. Therefore, the Sichuanoceratidae are considered to be a distinct group of the Orthocerida.

Genus SICHUANOCERAS Chang, 1962

Type species. *Sichuanoceras quizhouense* Chang, 1962, from the mid Silurian of Quizhou, China; by subsequent designation of Tsou (1966, p. 27).

Emended diagnosis. Straight slender conch with circular or subcircular cross-section; surface marked with transverse lines; sutures straight, transverse and with lobe on prosiphuncular side; siphuncle submarginal in position, about one-third of conch cross-section in diameter; septal necks suborthochoanitic; endosiphuncular deposits form massive rings that can fill nearly entire siphuncular segments; annuli positioned at septal perforation, nearly symmetrical in adoral and adapical directions, with subquadratic or rectangular shape in median section, typically more developed on side of siphuncle that is closer to conch margin; hyposeptal and episeptal deposits.

Remarks. The diagnosis of *Sichuanoceras* largely follows Tsou (1966, p. 27) who, however, failed to provide information on the shape of the septal necks and details of the endosiphuncular deposits, which are crucial for determination of the genus. Here I include these characters, which occur in the type species, based on his illustrations.

Occurrence. Mid Silurian–Early Devonian; China and Morocco.

Sichuanoceras zizense sp. nov.
Plate 14, figure 16

Derivation of name. From Qued Ziz, a large dry valley in southeastern Morocco, which crosses the Tafilalt in a north–south direction to the east of Taouz.

Holotype. MB.C.9509 (Pl. 14, fig. 16).

Type locality and horizon. Filon Douze section, bed OJ-I, Lochkovian.

Other material. 1 specimen, MB.C.9501, from the type locality and horizon.

Diagnosis. *Sichuanoceras* with slightly compressed cross-section and angle of expansion of *c.* 5–6 degrees; surface smooth or with very fine, widely spaced, transverse growth lines; septal distance *c.* 0.4 of conch cross-section; depth of septal curvature nearly equals septal distance; siphuncle eccentric with siphuncular segments slightly expanded; segments with width to height ratio of 0.8; diameter of septal perforation about one-fifth of conch cross-section; septal necks suborthochoanitic; endosiphuncular deposits strongly asymmetrical in dorsoventral direction, concentrated on ventral side of siphuncle.

Description. Conch surface of holotype poorly preserved but apparently smooth or with only very fine, widely spaced, growth lines. Total length of holotype 30 mm; conch height 33–36 mm (angle of expansion 8.4 degrees), with two complete chambers; adoral chamber 13 mm long; adapical chamber 15 mm long. Septal curvature of most adoral septum 11 mm. Septal perforation at most adoral septum 8.8 mm, which is 0.24 of conch-cross-section. Centre of septal perforation 16.4 mm from conch margin. Septal necks suborthochoanitic, slightly bent. Length of septal neck of most adoral septum 1.2 mm. Siphuncular segments slightly expanded within chambers, forming constriction at septal perforation. Maximum diameter of siphuncular segment at most adoral chamber 10.3 mm. Endosiphuncular deposits consist of rings positioned on ventral side of septal necks. Ratio between length of adoral half and adapical half of rings is 1.1. Contact surface between two succeeding annuli perpendicular to length axis of siphuncle. Growth surfaces of annuli reveal that they were initially elliptical or circular in longitudinal section but in later growth stages rectangular with nearly straight dorsal surfaces. At centre of each ring, at septal necks, narrow lamellae occur with a height of *c.* 0.3 of cross-section of septal perforation. Dorsal one-quarter of siphuncle free of deposits in holotype.

Second fragment 37 mm long; cross-section diameter 30–33 mm (angle of expansion 8.4 degrees); three complete chambers with lengths between 13 mm (adoral chamber) and 11 mm (adapical chamber). Septal curvature of most adoral septum 11 mm. Septal perforation at most adoral septum 6.8 mm, which is 0.21 of conch cross-section. Centre of septal perforation *c.* 15 mm from conch margin. Septal necks suborthochoanitic. Siphuncular segments slightly expanded within chambers, forming constriction at septal perforation. Maximum diameter of siphuncular segment at most adoral chamber 8.8 mm. Endosiphuncular deposits as in holotype.

Remarks. This species differs from all other species of *Sichuanoceras* in having a relatively narrow siphuncle that is only weakly expanded within the chambers and also positioned relatively close to the growth axis.

Family SPHAERORTHOCERATIDAE Ristedt, 1968

Genus SPHAERORTHOCERAS Ristedt, 1968

Type species. *Sphaerorthoceras beatum* Ristedt, 1968, from the Gorstian (Ludlow) of the Massif de Mouthoumet, Aude, France.

Diagnosis (after Ristedt 1978, p. 53). Orthocones with circular cross-section; shell surface smooth; sutures straight; siphuncle subcentral; septal necks orthochoanitic or suborthochoanitic; apex with spherical initial chamber and deep first constriction, without conspicuous second constriction.

Remarks. Serpagli and Gnoli (1977) remarked that *Sphaerorthoceras* is not sufficiently distinct from *Michelinoceras* Foerste, 1932, and thus questioned the value of the genus. However, *M. michelini* (Barrande, 1866), the type species, has elongate, slightly concave orthochoanitic septal necks and the apex shows only an inconspicuous first and second apical constriction (see Ristedt 1968, p. 245). Thus the diagnostic characters of *Michelinoceras* are very precisely known, even for the apical parts; its distinction from *Sphaerorthoceras* is, therefore, justified.

Occurrence. Silurian–Late Devonian; Carnic Alps (Austria), Czech Republic, Sardinia and Morocco.

Sphaerorthoceras sp. A
Plate 13, figure 20

Material. 1 apical fragment, MB.C.9510, from bed BS, Lochkovian, Filon Douze section.

Description. Apical fragment 0.9 mm long with adoral diameter of 0.6 mm. Conch slightly bent with circular cross-section and smooth surface. Spherical initial chamber 0.4 mm in diameter. Initial apical constriction has conch width of 0.5 mm at 0.35 mm from tip. Conch adoral of initial constriction expands with an apical angle of 22 degrees. Siphuncle central.

Remarks. The spherical initial chamber and smooth shell surface of these fragments are diagnostic of *Sphaerorthoceras*, but fragmentary nature and poor preservation of the material prevents specific determination.

Sphaerorthoceras sp. B
Plate 13, figure 1

Material. 1 apical fragment, MB.C.9511, from bed BS, Lochkovian, Filon Douze section.

Description. Apical fragment 0.9 mm long with adoral diameter of 0.4 mm. Conch straight with circular cross-section and smooth surface. Spherical initial chamber 0.25 mm in diameter. Initial apical constriction has conch width of 0.25 mm at 0.2 mm from tip. Initial constriction shallow. Conch adoral of initial constriction expands with an apical angle of 12 degrees. Siphuncle central.

Remarks. The fragment differs from *Sphaerorthoceras* sp. A in having a smaller initial chamber and a lower angle of expansion.

Genus AKROSPHAERORTHOCERAS Ristedt, 1968

Type species. *Akrosphaerorthoceras gregalis* Ristedt, 1968, from the Upper Silurian, cemetary in Fluminimaggiore, Sardinia, Italy.

Diagnosis (after Serpagli and Gnoli 1977, p. 175). Slender orthocones with circular cross-section; shell surface smooth in early growth stages, transversely undulating in later growth stages; sutures straight; siphuncle position eccentric; septal necks orthochoanitic; apex with conspicuous initial constriction and elongated initial chamber with acute, pointed tip; without conspicuous second constriction.

Occurrence. Wenlock–Lochkovian; Carnic Alps (Austria), Czech Republic, Sardinia and Morocco.

Akrosphaerorthoceras sp.
Plate 13, figures 13, 15

Material. 1 apical fragment, MB.C.9593, from bed OSP-II, Přídolí; 3 apical fragments, MB.C.9596, 9627, 10084, from bed BS, Lochkovian, Filon Douze section.

Description. MB.C.9593 is fragment of apical 4 mm with adoral diameter of 0.5 mm. Conch slightly bent. Conch surface with shallow undulations of *c.* 0.25 mm. Elongate, acute initial chamber has maximum cross-section diameter of 0.2 mm. Shallow, slightly oblique initial constriction occurs at 0.15 mm from tip. Conch section adoral of initial constriction slightly inflated, 0.4 mm in maximum cross-section diameter 0.75 mm from tip. Conch adoral of initial constriction until position of maximum inflation expands with apical angle of 18 degrees. Adoral of maximum inflation, conch nearly tubular.

Three specimens from bed BS show very little morphological variation. MB.C.9627 (Pl. 13, fig. 13) is representative of all three. It is a fragment of apical 1.5 mm with adoral diameter of 0.5 mm. Conch slightly bent; cross-section circular; siphuncle eccentric. Conch surface with shallow undulations of *c.* 0.25 mm. Elongate, acute initial chamber has maximum cross-section diameter of 0.25 mm. Shallow, slightly oblique initial constriction occurs 0.2 mm from tip. Conch section adoral of initial constriction slightly inflated, 0.5 mm in maximum cross-section diameter 1.25 mm from tip. Conch adoral of initial constriction to maximum inflation expands with apical angle of 14 degrees. Adoral of maximum inflation, conch nearly tubular.

Remarks. The acute, elongate initial chamber and slightly undulating shell surface of the fragments described above are diagnostic of *Akrosphaerorthoceras*. However, no species has been described with undulation starting directly at the tip. The fragmentary character and the poor preservation of the specimens prevents specific determination.

Genus HEMICOSMORTHOCERAS Ristedt, 1968

Type species. *Hemicosmorthoceras laterculum* Ristedt, 1968, from the basal limestone of the *Rhynchotrypa megaera* Beds, *Ozarkodina eosteinhornensis* conodont Biozone, Ludfordian (Ludlow) of Cellon, Plöckenpass, Carnian Alps, Austria.

Emended diagnosis. Slender, slightly compressed orthocones ornamented with transverse striae, undulations or smooth; sutures straight; siphuncle subcentral; septal necks orthochoanitic; apex with conspicuous first and second constriction, forming ovate or spherical initial chamber and strongly inflated conch section between apical constrictions; apical conch with shallow undulations, compressed cross-section; second constriction leading to slender, slowly widening conch with initially, clearly smaller cross-section diameter than conch section adapical of second constriction.

Remarks. Ristedt's (1968, p. 279) diagnosis is very short and provides only apical characters. Here, I include information on the shape of the septal necks and add some details on the characteristic shape of the apical conch based on the original illustration and description of the type species. *Hemicosmorthoceras* differs from other sphaerorthoceratids in having a depressed, strongly inflated apical conch section.

Occurrence. Ludlow–Lochkovian; Bohemia, Carnic Alps (Austria), Czech Republic, Morocco, Sardinia and Wales.

Hemicosmorthoceras aichae sp. nov.
Plate 13, figures 18–19

Derivation of name. In honour of Aicha, elder daughter of Filon Douze miner Lahcan Caraoi.

Holotype. MB.C.9586 (Pl. 13, figs 18–19).

Type locality and horizon. Filon Douze section, bed OJ-I, Lochkovian.

Other material. Known only from the holotype.

Diagnosis. *Hemicosmorthoceras* with strongly inflated conch section between apical constrictions with angle of expansion of *c.* 27 degrees; initial constriction occurs at second septum; second constriction leading to slender, slowly widening conch with initial conch width of 0.9 mm and angle of expansion of 2–3 degrees; conch section apical of second constriction bent with turning point at second constriction and at an angle of 20 degrees.

Description. Fragment 4.5 mm long, consisting of three sections: initial shell, inflated conch section, and shaft. Initial shell subspherical, 0.5 mm wide, separated from subsequent conch by conspicuous constriction 0.3 mm from tip. Adoral of initial constriction conch expands at an angle of 27 degrees towards width of 1.25 mm at 1.75 mm from tip, at which point cross-section depressed; width 1.5 mm; height 1.25 mm. At 1.75–2.25 mm from tip, conch width and height decrease strongly and cross-section becomes circular or slightly compressed. Inflated part of fragment slightly ornamented with transverse undulation. Distance between ridges of undulation *c.* 0.25 mm. Remaining part of conch smooth. Adoral of inflated conch section, shaft follows; this is 2.5 mm long and expands at angle of 2–3 degrees from 0.9 to 1 mm. Cross-section of shaft circular or slightly compressed. At second constriction conch bent, forming angle between initial shell and shaft of *c.* 20 degrees. Second septum occurs at position of initial constriction, tenth septum at second constriction *c.* 2.1 mm from tip.

Remarks. The conspicuous initial constriction, the broadly inflated conch section adoral of the initial constriction, and the smooth, slender shaft are diagnostic of *Hemicosmorthoceras*. However, the conspicuous curvature between initial shell and shaft and the ornamented inflated part of the apex are unique within the genus. The specimen described above is probably a complete juvenile conch with the last septum at the second constriction. However, it is possible that more adoral septa are not preserved.

Hemicosmorthoceras semimbricatum Gnoli, 1982
Plate 10, figures 2–3

1982 *Hemicosmorthoceras semimbricatum* Gnoli, p. 75, pl. 1, figs 1–3, text-fig. 2a–d.
1994 *Hemicosmorthoceras semimbricatum* Gnoli; Evans, p. 127.

Material. 1 specimen, MB.C.9530, from limestone nodules in bed OP-M, Ludfordian, Ludlow; 2 specimens, MB.C.10135–36, from bed OSP-II, Přídolí; 2 specimens, MB.C.9532–33, from bed PK; 4 specimens, MB.C. 9529, 10089–91, from bed CK; 80 specimens, MB.C.9534.1–80, from bed OJ-I, Lochkovian; Filon Douze section.

Emended diagnosis. Straight, slender orthocone with elliptically compressed cross-section; angle of expansion of conch height <5 degrees; ornamented with shallowly imbricated costae on prosiphuncular side; costae slope adorally and become indistinct on lateral and antisiphuncular sides, directly transverse on antisiphuncular side and curved towards apex on lateral side; sutures straight and slightly oblique; septal distance approximately equals conch cross-section diameter; septal perforation slightly eccentric, one-tenth of conch height in diameter; septal necks orthochoanitic, length similar to diameter of septal perforation. Endosiphuncular and cameral deposits not known; embryonic conch smooth.

Description. Largest specimen (MB.C.9529) is a fragment of body chamber 26 mm long; conch at adapical end (= position of last septum) 13 mm high and 11 mm wide; height at adoral end 15 mm. Cross-section compressed, slightly ovate with siphuncle in narrower half of section. Depth of septal curvature 3.4 mm. Septal perforation 5 mm from conch margin. On prosiphuncular side conch ornamented with shallowly imbricated ridges. Width of fine grooves between ridges *c.* 1.2 mm. Impression of the conch surface in MB.C.9532 reveals that costae curved towards apex on lateral side (Pl. 10, fig. 2).

Remarks. Gnoli's (1982, p. 76) diagnosis is modified above to include information on the adult; several references were made to the alleged small size of the adult of this species but no measurements were provided. Some of the Filon Douze specimens are clearly larger than that illustrated by Gnoli but are similar in all aspects of conch morphology. The species is easy to distinguish from other species of *Hemicosmorthoceras* by its oblique, imbricate costae that cover exclusively the prosiphuncular side of the conch.

Occurrence. Ludlow (Ludfordian)–Lochkovian; Morocco and Sardinia.

Genus PARASPHAERORTHOCERAS Ristedt, 1968

Type species. Parasphaerorthoceras accuratum Ristedt, 1968, from the *Ozarkodina eosteinhornenis* conodont Biozone, Ludfordian of Plöckenpass, Carnic Alps, Austria.

Diagnosis (after Ristedt 1968, p. 267). Orthocones with circular cross-section; shell surface undulating apically and smooth or striated in later growth stages; straight sutures; central or subcentral siphuncle; septal necks suborthochoanitic; apex with spherical initial chamber and deep first constriction; without conspicuous second constriction.

Remarks. Ristedt's (1968) diagnosis of *Parasphaerortho-ceras* is very short, indicating that it is a sphaerorticeratid with undulating apical growth stages. Additional characters of the type species are included here. Special emphasis is placed on the morphology of later growth stages. The type specimen is undulating apically, later develops smooth striae, and then becomes smooth in late growth stages. Serpagli and Gnoli (1977) found a

similar pattern in a number of specimens collected in Sardinia. The specimen MB.C.9517, assigned here to *Parasphaerorthoceras* sp. G *sensu* Ristedt 1968, also displays a loss of ornamentation with increasing size. Hence, this character is considered diagnostic of the genus.

Occurrence. Wenlock–late Middle Devonian; Carnic Alps (Austria), Czech Republic, Sardinia and Morocco.

Parasphaerorthoceras sp. B *sensu* Ristedt 1968
Plate 13, figure 21

1968   *Parasphaerorthoceras* sp. B; Ristedt, p. 269, pl. 4, fig. 3–3*a*, text-fig. 4.1*c*.
1977   *Parasphaerorthoceras* sp. B Ristedt; Serpagli and Gnoli, p. 169, pl. 4, fig. 9.

Material. 1 fragment, MB.C.9524, from bed BS, Lochkovian, Filon Douze section.

Description. Fragment of apical 1.5 mm of conch; adoral diameter 0.8 mm. Conch cross-section circular. Initial chamber elongate egg-shaped, 0.8 mm long, diameter at first septum 0.5 mm; similar at initial constriction. Inflated section of conch adoral of initial contriction only partly preserved but shows typical convex conch margins in lateral view. Surface ornamented with straight, directly transverse undulation; ten undulations occur within a distance of 1 mm.

Remarks. The egg-shaped initial chamber and low angle of expansion adoral of initial constriction are diagnostic of *Parasphaerorthoceras* sp. B *sensu* Ristedt 1968. The fragment described above shows these characters but is generally larger and the initial constriction is more conspicuous than in Ristedt's specimen. A satisfactory species designation must wait until better material is available.

Occurrence. Wenlock–Lochkovian; Morocco and Sardinia.

Parasphaerorthoceras sp. G *sensu* Ristedt 1968
Plate 13, figure 22

1968   *Parasphaerorthoceras* sp. G; Ristedt, p. 274, pl. 4, fig 4; pl. 5, fig. 3.

Material. 1 fragment, MB.C.9517, from nodule in bed PK, Lochkovian, Filon Douze section.

Description. An impression of apical 17 mm of conch; cross-section 1.7 mm at adoral end. Initial chamber 0.8 mm long, and at initial constriction 0.8 mm wide. Initial chamber subspherical and smooth. Subsequent conch section slightly inflated; maximum width 1.2 mm at 2.0 mm from tip. Adoral of *c.* 3 mm from tip conch grows with angle of expansion of 5 degrees. Apical part of impression ornamented with slightly oblique, acute undulations; seven undulations occur within a distance of 1 mm. Distinctiveness of undulation decreases towards adoral end of specimen. Adoral of *c.* 7–8 mm from tip, shell surface smooth.

Occurrence. Wenlock–Lochkovian; Morocco.

Remarks. Ristedt's (1968) single specimen of *P.* sp. G, collected from the Upper Silurian of the Moroccan Meseta, has a large initial chamber, relatively inconspicuous initial constriction and inflation adoral of this constriction, and slightly oblique undulation. The specimen described here must remain in open nomenclature until better material is available.

Parasphaerorthoceras sp. H *sensu* Ristedt 1968.
Plate 13, figure 4

1968 *Parasphaerorthoceras* sp. H; Ristedt, p. 275, pl. 4, fig. 9; pl. 5, fig. 7.
1977 *Parasphaerorthoceras* sp. H Ristedt; Serpagli and Gnoli, p. 169, pl. 4, figs 11–15.

Material. 1 apical fragment, MB.C.9514, from limestone nodules in bed OP-M, Ludfordian; 3 apical fragments, MB.C.9521–23, from bed BS, Lochkovian; Filon Douze section.

Description. MB.C.9514 is apical fragment 2.4 mm long; adapical diameter 0.7 mm; adoral diameter 1.3 mm. Initial chamber not preserved. Adapical end of fragment represents first septum and initial constriction. Conch cross-section slightly depressed. Conch margins convex in median section. Surface ornamented with straight, transverse undulations; seven undulations occur within a distance of 1 mm.

Three specimens from bed BS are complete, morphologically very similar, fragments of apical 2–3 mm of conchs. Initial chamber 0.6 mm long, subspherical. At initial constriction first septum occurs at conch height of 0.7 mm. Maximum inflation of subsequent conch section at 1.2 mm from tip where height is 1.2 mm. Conch cross-section depressed 2.5 mm from tip; width to height ratio 1.3. Surface ornamented with straight, directly transverse undulation; ten undulations occur at distance of 1 mm.

Remarks. The size and shape of the conch and the transverse undulation of the fragments are characteristic of *P.* sp. H *sensu* Ristedt 1968, as recognized by Serpagli and Gnoli (1977) who described details of the septal neck and later growth stages. The reason why they retained this form in open nomenclature is difficult to understand, but the poor preservation of the Filon Douze material does not justify the erection of a new species here.

Occurrence. Wenlock–Lochkovian; Morocco and Sardinia.

Genus PLAGIOSTOMOCERAS Teichert and Glenister, 1952

Type species. Orthoceras pleurotomum Barrande, 1866, from the *Colonograptus colonus* Biozone, Kopanina Formation, Gorstian (Ludlow) of Na Brekvici, Prague-Butovice, Bohemia, Czech Republic.

Diagnosis (after Teichert and Glenister 1952, p. 741, and Kolebaba 1999, p. 9). Slender orthocones with circular or compressed cross-section; aperture strongly oblique, sloping adapically from antisiphuncular to prosiphuncular side; aperture slopes more strongly on lateral sides and flattens near antisiphuncular and prosiphuncular sides; ornamented with growth lines or weak ridges parallel to aperture; sutures straight or slightly oblique, form shallow lateral lobes; septa moderately concave; siphuncle subcentral and siphuncular segments tubular; septal necks orthochoanitic; episeptal, hyposeptal and mural deposits occur.

Remarks. This genus differs from *Protobactrites* and *Bogoslovskya* in having a distinctively oblique, transverse ornamentation, but the apical characters of species assigned to it are poorly known. The only unquestionable apices of *Plagiostomoceras* were figured by Dzik (1984, pl. 26, figs 1–2, 5); they are similar to *Sphaerorthoceras beatum* Ristedt, 1968, in having an initial chamber with a slightly smaller cross-section diameter than the subsequent conch directly adoral of the initial constriction. It is questionable whether the apex figured by Dzik (1984, pl. 27, fig. 1) and assigned to *Plagiostomoceras* cf. *angustum* (Holzapfel, 1885) falls within the circumscription of the genus. The initial chamber of this apex is very short and resembles a calotte, and the cross-section of the conch directly adoral of the initial constriction is larger than that of the initial chamber.

Plagiostomoceras is similar to *Merocycloceras* in conch shape and ornamentation, but the apex of the latter differs in having several swollen first chambers. Because the apex of many species of *Plagiostomoceras* is not known, the possibility exists that several species conventionally assigned to this genus may be more correctly assigned to *Merocycloceras*. The adult growth stages of *Hemicosmorthoceras* are indistinguishable from those of *Plagiostomoceras* but it has very conspicuous early growth stages. I assign conch fragments similar in shape and ornamentation to *Hemicosmorthoceras*, *Merocycloceras* and *Plagiostomoceras* to the last of these genera when the apical shape is not known, or where it differs from known species of the first two.

Occurrence. Middle Ordovician–Early? Devonian; world-wide.

Plagiostomoceras pleurotomum? (Barrande, 1866)
Plate 10, figure 11

*1866–74 *Orthoceras pleurotomum* Barrande, pp. 412–413, 587–588, 599, pl. 9, figs 21–23; pl. 224, figs 12–14; pl. 296, figs 1–24; pl. 366, figs 1–4.
1952 *Plagiostomoceras pleurotomum* (Barrande); Teichert and Glenister, p. 741.
1962 *Plagiostomoceras pleurotomum* (Barrande); Balashov and Zhuravleva, pl. 11, fig. 1.
1999 *Pleurostomoceras pleurotomum* (Barrande); Kolebaba, p. 9, pl. 3, figs 1–8, text-fig. 12*d*.
2002 *Pleurostomoceras pleurotomum* (Barrande); Kolebaba, p. 184, text-fig. 2*a*.

Material. 1 specimen, MB.C.9798, from nodules in OP-M, Ludfordian, Filon Douze section.

Diagnosis. Plagiostomoceras with slender shell and elliptical, compressed cross-section; ornamented with sharp, obliquely transverse ridges and shallow, wide valleys; ridges imbricate with steep slope towards apex; ridges slope in adapical direction on prosiphuncular side, forming wide shallow lobe on prosiphuncular side and conspicuous saddle on antisiphuncular side; every second ridge spans only two-thirds of circumference, disappearing on antisiphuncular side; in juvenile shell all ridges reach around entire circumference; sutures oblique, parallel to ornamentation; septal distance 0.5–0.8 of conch height; septal necks long, orthochoanitic; septal neck length greater than diameter of septal perforation; diameter of septal perforation *c.* 0.1 of conch-cross-section; siphuncle eccentric.

Description. Fragment of phragmocone 24 mm long, dorsoventral diameter 5.7 mm. Conch ornamented with distinctive oblique ridges with shallow valleys in between. Every second ridge spans entire circumference, these spaced at *c.* 1–1.3 mm. Intercalated ridges occur at distance of 4 mm adoral of every second first order ridge. Distance between sutures 1.7 mm.

Remarks. The diagnosis of this species is derived from the original description and figures of Barrande (1860–70). MB.C.9798 displays the typical ornamentation of *P. pleurotomum*, but the internal characters are not known.

Occurrence. Ludfordian–Lochkovian; Czech Republic and Morocco.

Plagiostomoceras bifrons (Barrande, 1866) comb. nov.
Plate 10, figures 13, 17

*1866 *Orthoceras bifrons* Barrande, pl. 367, figs 1–31; pl. 442, figs 26–28.

Material. 2 specimens, MB.C.9944, 10088, from bed OJ-I; Lochkovian, Filon Douze section.

Diagnosis. Plagiostomoceras with slender shell and elliptical, compressed cross-section; angle of expansion 2–5 degrees; conch surface ornamented with sharp, obliquely transverse lirae and shallow, wide valleys; lirae imbricate with steep slope towards apex; they slope in adapical direction on prosiphuncular side, forming wide, shallow lobe on prosiphuncular side and conspicuous saddle on antisiphuncular side; of four or five lirae only two span entire circumference; the rest span only two-thirds of circumference, disappearing on antisiphuncular side; in juvenile shell all lirae reach around entire circumference; sutures oblique, parallel to ornamentation; septal distance 0.3–0.5 of conch height; septal necks long, orthochoanitic; septal neck length greater than diameter of septal perforation, which is *c.* 0.1 of conch-cross-section; siphuncle eccentric.

Description. Larger, well preserved fragment (MB.C.9944) 5 mm long, conch height 7.1–8.3 mm (angle of expansion 5 degrees), adapical conch width 6.4 mm. Conch elliptically compressed with width to height ratio of 0.9. Shell ornamented with imbricate transverse lirae. At distance of 5 mm 12 oblique lirae occur on prosiphuncular side, forming conspicuous saddle on antisiphuncular side; here very faint, distance between them on antisiphuncular side *c.* 1 mm (Pl. 10, fig. 13). Siphuncle eccentric. Diameter of septal perforation 0.7 mm at most adoral septum. Septal distance *c.* 2 mm.

Smaller, well-preserved specimen (MB.C.10088) 21 mm long, with cross-section height 2.3–4.3 mm, adoral width 4 mm, represents juvenile growth stages with characteristic obliquely transverse lirae spanning entire circumference (Pl. 10, fig. 17). Eight lirae occur at distance of 3 mm. Septal distance 1.6 mm at adapical end of specimen.

Remarks. The diagnosis was compiled from the original description and figures of Barrande (1866). *Plagiostomoceras bifrons* resembles *P. pleurotomum* in general shell shape and ornamentation. However, in the latter species every second ridge spans the entire circumference and the angle of expansion is lower.

Occurrence. Lochkovian; Czech Republic and Morocco.

Plagiostomoceras culter (Barrande, 1866)
Plate 10, figures 9, 14, 18

*1866–70 *Orthoceras culter* Barrande, pl. 239, figs 9–11, pl. 334, figs 5–22; pl. 442, figs 7–8.
1874 *Orthoceras culter* Barrande, p. 626.
1991 *Plagiostomoceras culter* (Barrande); Gnoli and Serpagli, pp. 190, 194, pl. 3, fig. 6.
1999 *Plagiostomoceras culter* (Barrande); Histon and Gnoli, pp. 384, 388, tables 1–2.

Material. 1 fragment, MB.C.9947, from OP-K, Ludfordian; more than 100 specimens from bed PK, Lochkovian, Filon Douze section.

Diagnosis. *Plagiostomoceras* with slender, nearly smooth shell; compressed cross-section highly variable, from elliptical, egg-shaped to rounded triangular; ratio of conch width to height 0.65–0.75; ornamented with faint, irregularly spaced, oblique growth lines forming conspicuous saddle on prosiphuncular side; sutures straight or very slightly oblique, forming conspicuous lateral lobe; septal spacing close at 0.2–0.4 of conch height; septal perforation <0.1 of conch height in diameter; siphuncle eccentric.

Description. Specimen MB.C.9947, a fragment of three chambers with adapical part of body chamber, is 34 mm long; conch height 19–21 mm; adoral width 19 mm. Height of each of the three chambers 2 mm. Sutures form shallow lateral lobe. Shell thickness at base of body chamber 0.3 mm. Faint thickening of shell of body chamber at distance of *c.* 16 mm from last septum forming shallow undulation on mould of chamber. Faint, irregularly spaced growth lines form conspicuous saddle on one side of conch (Pl. 10, fig. 9).

MB.C.9943, a fragment of five chambers with adapical part of body chamber, is 40 mm long; conch height 20–23 mm; adoral width 19 mm (Pl. 10, fig. 18). Height of chambers 1.8–2.5 mm. Most adoral two chambers narrowest. Sutures form shallow lateral lobe. Length of body chamber 32 mm. Conch margins in median section slightly convex. At adoral end of fragment of body chamber shallow constriction occurs.

MB.C.9953 is fragment of body chamber and one septum, showing siphuncle, which is eccentric, at conch height of 14.6 mm; centre of septal perforation 6.1 mm from conch margin. Diameter of septal perforation 1.3 mm.

Remarks. The diagnosis of this species is based on the original figures of Barrande (1860–70). Compared to his specimens, those from the Filon Douze section have a less variable cross-section, being invariably compressed elliptical; in some specimens it is egg-shaped with the siphuncle on the narrower side of conch.

Occurrence. Ludfordian–Lochkovian; Czech Republic, Morocco and Sardinia.

Plagiostomoceras lategruenwaldti sp. nov.
Plate 10, figure 12

Derivation of name. Latin, *late*, wide, referring to the transverse ornamentation, which is similar to that of *P. gruenwaldti* (Barrande, 1868) but more widely spaced.

Holotype. MB.C.9954 (Pl. 10, fig. 12).

Type locality and horizon. Filon Douze section, bed OJ-I, Lochkovian.

Other material. 1 specimen, MB.C.9955, from the type locality and horizon.

Diagnosis. *Plagiostomoceras* with elliptically compressed cross-section and straight conch; angle of expansion *c.* 7 degrees; conch width to height ratio *c.* 0.8; ornamented with narrow, transverse undulation and faint longitudinal striae; undulations obliquely transverse, forming conspicuous lateral saddle, irregularly spaced, about six undulations occur within distance similar to conch height; sutures slightly oblique, sloping parallel to transverse ornamentation; septal distance 0.5 of conch cross-section; siphuncle eccentric.

Description. Holotype 14.5 mm long; conch height 6.8–8.6 mm; adoral width 7 mm; angle of expansion 7 degrees. Surface ornamented with irregularly spaced, narrow undulations and fine, shallow longitudinal striae. Striae between undulations highest adapical of each undulation, giving shell an imbricated appearance. At distance of 10 mm seven oblique, transverse undulations, sloping in adoral direction on prosiphuncular side, form distinctive lateral saddle. Sutures slightly oblique, sloping in similar direction to transverse ornamentation. Septal distance *c.* 5 mm. Siphuncle eccentric.

Remarks. *Plagiostomoceras gruenwaldti* is ornamented with irregularly spaced narrow undulations that are more closely spaced and less conspicuous than in *P. lategruenwaldti*.

Plagiostomoceras reticulatum sp. nov.
Plate 10, figures 7–8

Derivation of name. Latin, *reticulatus*, net-like, referring to the ornamentation of this species.

Holotype. MB.C.10116 (Pl. 10, figs 7–8).

Type locality and horizon. Filon Douze section, bed OJ-I, Lochkovian.

Other material. 8 specimens, MB.C.9638, 10110–15, 10131, from the type locality and horizon.

Diagnosis. *Plagiostomoceras* with slightly compressed cross-section and slightly bent, slender conch; angle of expansion *c.* 3 degrees; ornamented with transverse ridges, sloping slightly in adoral direction on convex side of curvature of growth axis, forming shallow lateral lobe and distinctive saddle on this side of conch curvature; about ten ridges within distance similar to conch cross-section; faint longitudinal striae form shallow reticulate pattern; sutures slightly oblique, sloping parallel to transverse ridges; septal distance 0.4 of conch cross-section; siphuncle subcentral.

Description. Holotype 14 mm long; conch height 5.7–6.3 mm; adoral width 5.6 mm; angle of expansion 2.5 degrees. Ten transverse ridges occur within 5 mm. Ridges slightly oblique, sloping in adoral direction on convex side of curvature of growth axis, forming shallow lateral lobe and distinctive saddle on this side of conch curvature. Faint, shallow, longitudinal striae cross transverse ridges. Distance between longitudinal striae *c.* 0.3 mm. Siphuncle nearly central. Diameter of septal perforation at most adapical septum 0.4 mm. Distance between septa *c.* 2 mm. Largest specimen (MB.C.10110) 14 mm long, cross-section height 9–9.5 mm; angle of expansion 2 degrees, ratio of conch width to height 0.95. Variation of ornamentation and conch shape very low in eight specimens known.

Remarks. This species resembles *Orthoceras placidum* Barrande, 1868 (species combined *Plagiostomoceras* below: see 'List of new combinations'), which is similarly ornamented with longitudinal striae, but the transverse ridges are clearly more oblique and possess conspicuous lateral lobes.

Order BACTRITIDA Shimansky, 1951
Family BACTRITIDAE Hyatt, 1884

Genus BACTRITES Sandberger, 1843

Type species. *Bactrites subconicus* Sandberger, 1843, from the Wissenbach Slate, Eifelian of Wissenbach, Germany.

Diagnosis (after Erben 1964, p. K501). Slender orthocones with nearly circular to broadly oval cross-section; sutures almost straight with small ventral lobe, on lateral sides rectiradiate or prorsiradiate; growth lines form broad, shallow ventral sinus and low dorsal saddle.

Occurrence. Pragian?–Permian; world-wide.

Bactrites gracile (Blumenbach, 1803)
Plate 14, figure 12

```
 * 1803   Orthoceras gracile Blumenbach, pl. 2, fig. 6.
   2005   Bactrites gracile (Blumenbach); Kröger et al.,
             pp. 331, 339, fig. 1A [with synonymy].
```

Material. 3 specimens, MB.C.9538–39, 9545, from bed F, *P. partitus* conodont Biozone, early Eifelian, Filon Douze section.

Diagnosis (from Kröger *et al.* 2005, p. 339). Smooth, straight or very slightly bent *Bactrites* with circular or compressed cross-section; sutures straight, slightly oblique; septal spacing variable with approximately two septa at distance similar to cross-section; initial chamber ovate, smooth, with circular cross-section, *c.* 0.9 mm wide and 1.2 mm long; ovate shape of initial chamber highly variable.

Description. MB.C.9945 51 mm long; dorsoventral diameter 13.5 mm; conch width 12.5 mm with ten chambers. Cross-section of fragment compressed oval, increases from 9.2 to 13.5 mm in diameter. Conch surface smooth. Septal perforation 1.4 mm wide at conch diameter of 12 mm. Septal necks 1.7 mm long and funnel-like.

Occurrence. Late Emsian–Givetian; Germany, Morocco, UK and USA.

Bactrites? sp.
Plate 13, figure 8

Material. 1 apical fragment, MB.C.5764, from bed KMO-III, Pragian, Filon Douze section.

Description. Fragment 2.2 mm long, adoral diameter 0.5 mm; very slightly bent. Initial 0.75 mm slightly inflated with maximum diameter of 0.45 mm. Shallow constriction adoral of inflated part at distance 0.75 mm from tip where conch diameter is 0.4 mm. Interval adoral of constriction grows with angle of expansion of 4 degrees. First chamber 0.55 mm high, second and third chambers 0.2 mm high. Subsequent chambers 0.25, 0.35 and 0.5 mm high respectively, measured from apical direction adorally.

Remarks. The entire specimen was ground successively parallel to the longitudinal axis to locate the position of the siphuncle. During this process no siphuncle was found although the individual septa were completely preserved; there were no septal perforations in a central or subcentral position in the specimen.

During the preparation of numerous median sections of orthocones it was often extremely difficult to locate the siphuncular perforation of bactritoids because of their marginal position and small diameter. It is concluded, therefore, that the specimen is probably the apex of a bactritoid and that the marginal septal perforation could not be seen during the grinding process.

Genus DEVONOBACTRITES Shimansky, 1962

Type species. *Orthoceratites obliquiseptatum* Sandberger and Sandberger, 1852, from the Wissenbach Slate, Eifelian of Wissenbach, Germany.

Emended diagnosis. Smooth, straight or very slightly bent bactritoids with circular or compressed cross-section; sutures oblique, sloping in adoral direction on prosiphuncular side; septal spacing close with *c.* 3–4 septa at distance similar to conch cross-section; initial chamber elongate egg-shaped with acute tip and compressed cross-section.

Remarks. Devonobactrites was erected with only the briefest of diagnoses: 'Differs from *Bactrites* in having a more sloped suture and shorter chambers' (Shimansky 1962, p. 236). Erben (1964, p. K503) questioned the 'independence' of the genus because of the high variability of the diagnostic characters. Ignoring the discussion of the concept of 'independence' of a genus, I accept the genus and include the apical characters in the diagnosis. The apex of *Devonobactrites obliqueseptatus* differs considerably in size and shape from that of *Bactrites gracile*.

Occurrence. Early Emsian (Zlíchovian)–early Eifelian; Germany and Morocco.

Devonobactrites obliqueseptatus (Sandberger and Sandberger, 1852)
Plate 13, figures 9–10; Plate 14, figures 6–9

* 1852 *Orthoceras obliqueseptatum* Sandberger and Sandberger, p. 160, pl. 18, figs 2, 2a–e.
  1876 *Orthoceras oblique-septatum* Sandberger and Sandberger; Maurer, p. 829.
  1878 *Orthoceras obliqueseptatum* Sandberger and Sandberger; Kayser, p. 81, pl. 20, fig. 18.
  1962 *Devonobactrites obliquiseptatus* (Sandberger and Sandberger); Shimansky, p. 263, pl.1, fig. 6.
  1964 *Devonobactrites obliqueseptatus* (Sandberger and Sandberger); Erben, p. K503, fig. 361, 2a–c.
  2005 *Bactrites* sp. C; Kröger *et al.*, p. 334, figs 4A–B, 5A.
  2008 *Devonobactrites obliquiseptatus* (Sandberger and Sandberger); Klug *et al.*, pp. 48–49, pl. 6, figs 11–12; pl. 7, figs 1–24.

Material. 467 fragments of phragmocone and body chamber, MB.C.9542–47, 9568; 19 apices, MB.C.9548–49, 9550.1–17, from bed EF, Zlíchovian, Emsian, Filon Douze section.

Emended diagnosis. Devonobactrites with compressed cross-section, close septal spacing with approximately four septa at distance similar to cross-section; angle of expansion *c.* 5 degrees; cross-section of base of adult body chamber *c.* 5 mm; length of adult body chamber *c.* 20 mm; apex elongate egg-shaped with acute tip; faint constriction and first septum at *c.* 0.6–0.9 mm from tip with diameter of 0.4 mm; diameter of initial conch 0.3–0.4 mm.

Description. Conch nearly smooth with faint, oblique undulations in late growth stages, which form lobe on prosiphuncular side. Faint, oblique growth lines slope adapically on prosiphuncular side parallel to the undulations. Conch grows constantly with angle of expansion averaging 5 degrees (360 specimens measured; standard deviation 2.3 degrees). Cross-section ovate-compressed with average width to height ratio of 0.89 (423 specimens measured; standard deviation 0.12). Mature septal crowding occurs at cross-section diameter of 3–4 mm. Mature body chamber *c.*

18 mm long, with maximum cross-section diameter of chamber *c.* 7 mm. Septa slightly oblique, sloping adorally on prosiphuncular side of conch. Obliquity of septa varies between specimens. Septal spacing 0.2–0.25 of conch cross-section (Pl. 14, fig. 7). In early juvenile growth stages, angle of expansion very low and septal spacing more than 0.7 of conch cross-section. Sutures oblique, forming shallow lateral lobes and broad saddle on antisiphuncular side. Siphuncle marginal and very narrow.

Apex (Pl. 13, figs 9–10) 0.6–0.84 mm long and 0.32–0.4 mm in diameter (15 specimens measured). Diameter at initial constriction 0.32–0.36 mm (13 specimens measured). Conch adoral of initial chamber nearly straight with very low angle of expansion and wide chamber distances. Septal perforation at initial septum marginal. Cross-section of apical conch slightly compressed.

Remarks. The large quantity of material collected from the Filon Douze section provides information on apical and adult characters and intraspecific variation. All these data are now included within the emendation of *D. obliqueseptatus.*

Occurrence. Early Emsian (Zlíchovian)–early Eifelian; Germany and Morocco.

Devonobactrites emsiense sp. nov.
Plate 16, figures 1–2

Derivation of name. From Emsian, with reference to the type horizon.

Holotype. MB.C.9531 (Pl. 16, figs 1–2).

Type locality and horizon. Filon Douze section, bed KMO-V, Dalejian, late Emsian.

Other material. 3 specimens, MB.C.9535–37, from the *Erbenoceras* Limestone, Zlíchovian, early Emsian, Filon Douze section.

Diagnosis. Devonobactrites with very slightly bent conch; angle of expansion *c.* 3 degrees; cross-section compressed elliptical, with conch width to height ratio of *c.* 0.8; three septa at distance similar to conch cross-section; suture straight or with very shallow lateral lobes, slightly oblique, sloping adorally on prosiphuncular side.

Description. Holotype 40 mm long, cross-section compressed elliptical, with conch height 11–13 mm and adoral conch width 11.6 mm. Conch surface not preserved. Fragment has ten slightly oblique septa that slope in adoral direction on prosiphuncular side. Sutures straight or with shallow lateral lobe.

MB.C.9535 is single chamber of phragmocone 3.9 mm long, 13.5 mm high, 12 mm wide. Diameter of siphuncle 1.45 mm. MB.C.9536 is 13.5 mm in maximum diameter and displays three chambers with depth of 4.5 mm, and smooth surface of outer shell.

Remarks. Devonobactrites emsiense resembles *Bactrites au-savensis* Steininger, 1853, from the Frasnian of Germany, the cross-section of which, however, is less compressed and the sutures have conspicuous lateral lobes. It differs from *D. obliqueseptatus* in having a less compressed cross-section, slightly wider septal spacing and a larger adult size.

Devonobactrites sp.
Plate 14, figure 15

Material. 3 fragments of the body chamber, MB.C.10195–97, from bed EF, Zlichovian, early Emsian, Filon Douze section.

Description. MB.C.10195 is 11.6 mm long; cross-section height 2.48–2.73 mm. MB.C.10196 is 18.6 mm long; cross-section height 3.47–4.25 mm. MB.C.10197 is 20.4 mm long; cross-section height 6.9–7.6 mm. Conch of all three specimens compressed; conch height to width ratio 0.89–0.98 (smallest–largest fragment). Sutures slightly oblique.

Remarks. These fragments differ from *D. obliqueseptatus* in having a significantly lower angle of expansion. It is impossible to diagnose a new species on the basis of this material.

Genus LOBOBACTRITES Schindewolf, 1932

Type species. Bactrites ellipticus Frech, 1897, from the Wissenbach Slate, Eifelian of Wissenbach, Germany.

Diagnosis (after Erben 1964, p. K501; Schindewolf 1933). Slender bactritoids with narrowly oval cross-section; some forms have flattened lateral sides and rarely dorsal carina; sutures with small ventral lobe, well-developed lateral lobes and dorsal saddle; growth lines similar to those of *Bactrites* but more strongly rursiradiate, dorsal saddle more prominent and commonly pointed; initial chamber elongate-ovate with circular cross-section, width of 0.9 mm and length of 1.5 mm.

Occurrence. Emsian–Late Devonian; Australia; China; Germany, Morocco, North America and Russia.

Lobobactrites ellipticus (Frech, 1897)
Plate 14, figures 10–11

1852 *Bactrites carinatus* Münst. sp.; Sandberger and Sandberger, p. 129, pl. 17, fig. 3a–d, f–n (*non* pl. 17, fig. 3e).
* 1897 *Bactrites ellipticus* Frech, pl. 30a, fig. 7a–c.
1932 *Lobobactrites ellipticus* (Frech); Schindewolf, p. 174, text-fig. 5c.
1933 *Lobobactrites ellipticus* (Frech); Schindewolf, p. 73, pl. 3, figs 1, 7; pl. 4, fig. 8; text-fig. 15.

1960 *Lobobactrites ellipticus* (Frech); Erben, pp. 17, 33, pl. 2, figs 2–3; text-fig. 5.
1984 *Lobobactrites ellipticus* (Frech); Dzik, p. 105, text-figs 39–40.

Material. 1 specimen, MB.C.9580, from bed F, *P. partitus* conodont Biozone, Eifelian, Filon Douze section.

Diagnosis. Lobobactrites with narrowly compressed oval cross-section; sutures with small ventral lobe, well-developed lateral lobes and dorsal saddle; surface smooth or with faint growth lines.

Description. Fragment of phragmocone 39 mm long; conch height 6.0–6.2 mm. Cross-section elliptically compressed; conch width to height ratio 0.65; nine chambers with sutures that have conspicious lateral lobes; septal spacing highly variable.

Remarks. When erected, *L. ellipticus* was not described thoroughly. In the caption of a plate in the atlas of a textbook (Frech 1897) there is only a short note that the siphuncle is marginal. Schindewolf (1932, 1933) described and defined *Lobobactrites* but failed to provide a detailed description of the species. Here, I regard *L. ellipticus* as a *Lobobactrites* that shows the general characters of the genus but lacks flattened lateral sides and carinae and has a smooth shell.

Occurrence. Wissenbach Slate, Eifelian; Germany, and *P. partitus* conodont Biozone, Eifelian, Morocco.

LIST OF NEW COMBINATIONS

Some of the species noted above under 'Other species included' in a genus belong to different genera: they are referred to the genera to which they were originally assigned by their authors. Three are mentioned in different contexts and some are included in discussions. These taxa are formally combined with the genera concerned here. For the sake of completeness, the new combination of *Plagiostomoceras bifrons* formally recorded above is included in this list.

Ventrobalashovia altajense (Zhuravleva, 1979) comb. nov.
Tripleuroceras altajense Zhuravleva, 1979, p. 305, pl. 6, figs 3–5.

Ankyloceras sardoum (Serpagli and Gnoli, 1977) comb. nov.
Galtoceras? *sardoum* Serpagli and Gnoli, 1977, p. 192, pl. 9, fig. 1a–c.

Orthorizoceras sinkovense (Balashov, *in* Balashov and Kiselev 1968) comb. nov.
Metarizoceras sinkovense Balashov, *in* Balashov and Kiselev 1968, p. 15, pl. 3, fig. 4.

Subormoceras dobrovljanense (Kiselev, *in* Balashov and Kiselev 1968) comb. nov.
Ormoceras dobrovljanense Kiselev, *in* Balashov and Kiselev, p. 24, pl. 3, fig. 2a–b.

Subormoceras rashkovense (Kiselev, *in* Balashov and Kiselev 1968) comb. nov.
Ormoceras rashkovense Kiselev, *in* Balashov and Kiselev 1968, p. 23, pl. 7, figs 1–3.

Subormoceras skalaense (Kiselev, *in* Balashov and Kiselev 1968) comb. nov.
Ormoceras skalaense Kiselev, *in* Balashov and Kiselev 1968, p. 24, pl. 7, figs 4–5.

Cancellspyroceras loricatum (Barrande, 1868) comb. nov.
Orthoceras loricatum Barrande, 1868, p. 277 (1874), pl. 275, figs 10–13, pl. 424, figs 5–8.

Infundibuloceras mira (Zhuravleva, 1978) comb. nov.
Bogoslovskya mira Zhuravleva, 1978, p. 59, pl. 3, fig. 13.

Infundibuloceras komiense (Zhuravleva, 1978) comb. nov.
Sinoceras komiense Zhuravleva, 1978, p. 49, pl. 1, figs 11–12.

Tibichoanoceras abditum (Kiselev, *in* Balashov and Kiselev 1968) comb. nov.
Michelinoceras abditum Kiselev, *in* Balashov and Kiselev 1968, p. 10, pl. 1, fig. 4.

Tibichoanoceras riphaeum (Zhuravleva, 1978) comb. nov.
Sinoceras riphaeum Zhuravleva, 1978, p. 47, pl. 1, figs 3–7.

Adiagoceras subnotatum (Barrande, 1868) comb. nov.
Orthoceras subnotatum Barrande, 1868, p. 386 (1874), pl. 307, figs 5–8.

Angeisonoceras consocium (Barrande, 1870) comb. nov.
Orthoceras consocium Barrande, 1870, p. 452 (1874), pl. 372, figs 15–17; pl. 373, figs 8–9; pl. 375, figs 16–20; pl. 387, figs 4–6.

Angeisonoceras davidsoni (Barrande, 1870) comb. nov.
Orthoceras davidsoni Barrande, 1870, p, 496 (1874), pl. 391, figs 5–6, 8; pl. 392, figs 1–4, 8–11; pl. 393, figs 5–8; pl. 445, figs 3–5.

Angeisonoceras festinans (Barrande 1870) comb. nov.
Orthoceras festinans Barrande 1870, p. 451 (1874), pl. 373, figs 10–13.

Angeisonoceras imminutum (Barrande, 1870) comb. nov.
Orthoceras imminutum Barrande, 1870, p. 375 (1874), pl. 373, fig. 14; pl. 375, figs 21–24.

Angeisonoceras inchoatum (Barrande, 1868) comb. nov.
Orthoceras inchoatum Barrande, 1868, p. 328 (1874), pl. 209, figs 8–12; pls 368–369.

Angeisonoceras libens (Barrande, 1870) comb. nov.
Orthoceras libens Barrande, 1870, p. 452 (1874), pl. 387, figs 7–9.

Angeisonoceras limatum (Barrande 1870) comb. nov.
Orthoceras limatum Barrande 1870, p. 375 (1874), pl. 375, figs 1–4.

Angeisonoceras squamatulum (Barrande, 1868) comb. nov.
Orthoceras squamatulum Barrande, 1868, p. 455 (1874), pl. 302, figs 8–13.

Plagiostomoceras bifrons (Barrande, 1866) comb. nov.
Orthoceras bifrons Barrande, 1866, p. 398 (1874), pl. 367, figs 1–31; pl. 442, figs 26–28.

Plagiostomoceras placidum (Barrande, 1868) comb. nov.
Orthoceras placidum Barrande, 1868, p. 410 (1874), pl. 298, figs 18–35; pl. 367, figs 19–21.

CONCLUSIONS

The Ludfordian–Lochkovian section at Filon Douze is composed of several hundred metres of thick silty claystones that are intercalated by three horizons containing proximal tempestites and sandstones with a low diversity bivalve-orthocerid fauna. In contrast, the Pragian–Eifelian section is composed of an alternation of silty marls and nodular crinoid-dacryonoconarid limestones that contain a rich benthos of trilobites, rugose corals and gastropods.

EXPLANATION OF PLATE 16

Figs 1–2. *Devonobactrites emsiense* sp. nov.; MB.C.9531, holotype; bed KMO-V, Dalejian, Emsian. 1, lateral view; note the oblique septation. 2, ventral view, slightly weathered, showing position and shape of septal perforations.

Fig. 3. Orthoceridae gen. et sp. indet.; MB.C.9913; bed KMO-V, Daleijan, Emsian; median section.

Figs 4, 7. *Arionoceras* aff. *canonicum* (Meneghini, 1857); MB.C.10229; bed OP-M, Ludfordian, Ludlow. 4, lateral view. 7, adapical view; note comparatively strongly compressed conch.

Figs 5–6. *Arionoceras capillosum* (Barrande, 1867); MB.C.10206; bed BS, Lochkovian. 5, lateral view. 6, detail of lateral view; scale bar represents 1 mm.

Fig. 8. *Deiroceras hollardi* sp. nov.; MB.C.9630; *Deiroceras* Limestone, Zlíchovian, Emsian; detail of conch surface showing transverse lirae; scale bar represents 1 mm.

Figs 9–11. *Pseudendoplectoceras lahcani* gen. et sp. nov.; MB.C. 10184, holotype; bed KMO-II, Pragian. 9, median section of entire specimen. 10–11, showing the structure of the siphuncle and septal neck; note the recumbent septal necks and the endosiphuncular deposits; scale bar represents 1 mm.

All specimens from the Filon Douze section and × 2 except where indicated by scale bar.

PLATE 16

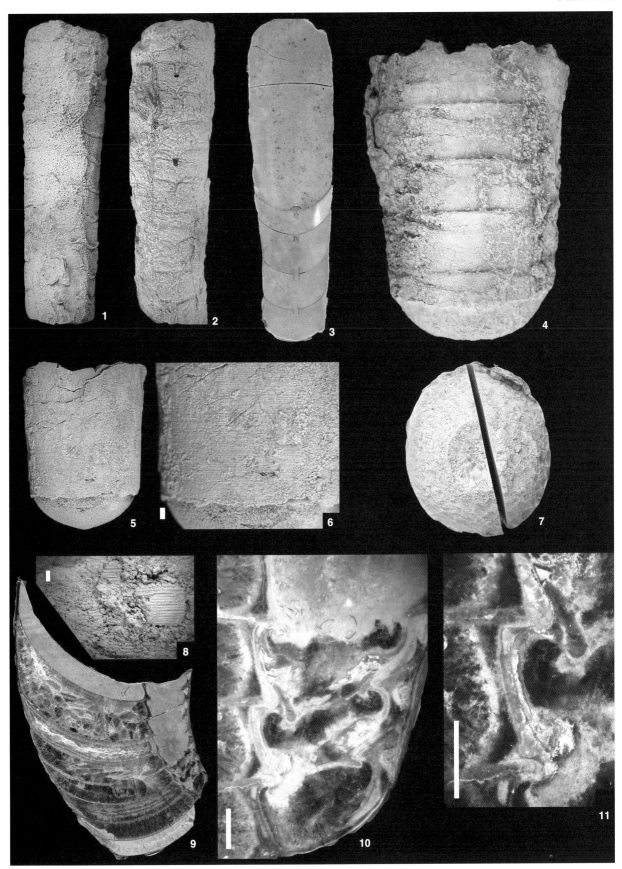

KRÖGER, nautiloids

In addition, condensed cephalopod limestones occur in the Emsian and Eifelian that commonly contain superficially corroded shell fragments. At the base of the post-Lochkovian limestone layers, there is often a horizon containing masses of dacryoconarids and large bivalves such as *Panenka*.

Through the entire section no general tendency of increasing or decreasing depth of deposition can be determined. However, several depositional cycles occur. Some of these contain sandstones at their base (e.g. at top of the *Temperoceras* limestone and *Jovellania* limestone successions), some display basal crinoidal-packstones (e.g. at the top of the *Scyphocrinites* limestone succession), others contain winnowed crinoid-dacryonoconarid packstones at the base (e.g. bed K2, *Deiroceras* Limestone). Transgressive peaks are marked by horizons with phosphatic nodules (bed PK) or limonitic concretions (e.g. bed EF) in the siltstones, or by omission horizons at the top of crinoid-dacryonoconarid pack- and wackestones (e.g. top of the *Deiroceras* Limestone). The condensed cephalopod limestones of the *Erbenoceras* Limestones, of beds PI and PII, are interpreted in terms of relatively shallow depositional settings close to, or above, storm wave-base (compare Wendt *et al.* 1984).

A major facies change at the top of the *Jovellania* Limestones from dark tempestitic facies towards light-coloured crinoid-dacryonoconarid limestones to marl alternations is interpreted to coincide with the *pesavis* Event in the latest Lochkovian. This facies change does not reflect a lasting sea-level drop at Filon Douze. The post-Lochkovian sediments are considered to reflect on average depths of deposition similar to those of the pre-Pragian sediments. A dramatically increased richness in the benthic fauna indicates strongly enhanced living conditions on the sea-floor or close to it in the water column in post-Lochkovian times. Similar facies and faunal changes have been reported from several other localities in Morocco, Bohemia, France, Spain and Sardinia (e.g. Alberti 1980; Kříž 1999; Chlupáč *et al.* 2000; Manda 2001; Robardet and Gutiérrez-Marco 2004). The latest Lochkovian *Jovellania* Limestones mark the last occurrences of several Arionoceratidae and Sphaerorthoceratidae, which are characteristic of Silurian and earliest Devonian deposits.

The prominent innovation of the cephalopod fauna occurred after an interval of 2–3 myr following this major facies change in the late Pragian–early Zlíchovian. During this period bactritoids and ammonoids appeared nearly simultaneously. In the late Zlíchovian the first appearance of *Centroceras* marked the beginning of the success of the Nautilitida, the coiled forms of which are very common in the late Palaeozoic. At Filon Douze this landmark change is not accompanied by a general increase in cephalopod richness. However, whereas in the pre-Pragian strata Oncocerida, Discosorida and Lituitida are rare

faunal components, these forms become more diverse in higher parts of the section. This trend is reflected by an increasing taxonomic distinctness during the time interval concerned. However, a supra-regional comparison of this pattern is currently impossible because of the lack of data for other regions.

Acknowledgements. This work was financially supported by the Deutsche Forschungsgemeinschaft, DFG Grant KR 2095/2. I am indebted to C. Klug (Zürich, Switzerland), S. Lubeseder (Leicester, UK), and A. El Hassani (Rabat, Morocco) for help, advice and discussion of the stratigraphical part of the paper. I am grateful to A. El Hassani for logistical support of the field trips. Manina Lintzmeyer and Marek Damm (Berlin, Germany), and Mark Simon (Hannover, Germany) helped in the preparation of the median sections and some of the illustrations. I thank Elke Sievert for preparing the hand drawings. I am grateful to the hospitality of the family of Filon Douze miner L. Caraoui (Hassilbid, Tafilalt). I thank M. Stenzel for mechanical preparation of the apices and S. Kiel (Leeds, UK) for joining me on my first field trip to Morocco in the spring of 2003. I am grateful to E. Shubert (London, UK) for initial linguistic review of the manuscript. I am indebted to P. D. Lane (Keele, UK) for considerable editorial and scientific advice during preparation of the revised version of this paper, and to D. J. Batten (Manchester, UK) for his editorial work. The paper is a contribution to IGCP Project 497, 'The Rheic ocean: its origin, evolution and correlatives'.

REFERENCES

AGASSIZ, J. L. R. 1847. *Nomenclator Zoologicus. Fasciculus XII. Indicem Universalem. Nomenclatoris Zoologici Index Universalis, continens nomina systematica classicum, ordinum, familiarum et genera animalium omnium, tam vivetum quam fossilium, secundum ordinem alphabeticum unicum disposita, adjectis homonymiis plantarum, nec non variis adnotationibus et emendationibus.* Vol. 7, Jent et Gassmann, Oloduri, 393 pp.

ALBERTI, G. K. B. 1969. Trilobiten des jüngeren Siluriums sowie des Unter- und Mitteldevons. I. *Abhandlungen der Senckenbergischen Naturforschenden Gesellschaft,* **520**, 1–692.

—— 1980. Neue Daten zur Grenze Unter-/Mittel-Devon, vornehmlich aufgrund der Tentaculiten und Trilobiten im Tafilalt (SE-Marokko). *Neues Jahrbuch für Geologie und Paläontologie, Monatshefte,* **1980**, 581–594.

BABIN, C. R. 1966. *Mollusques bivalves et céphalopodes du Paléozoïque armoricain. Étude systématique. Essai sur la phylogénie des Bivalves. Esquisse paléoécologique.* Imprimérie commerciale et administrative, Brest, 470 pp., 18 pls.

——DEUNFF, J., MÉLOU, M., PARIS, F., PELHATE, A., PLUSQUELLEC, Y. and RACHEBOEUF, P. R. 1979. La coupe de Porzar Vouden (Přídolí de la presqu'ile de Crozon) Massif Armoricain. *Palaeontographica, A,* **164**, 52–84.

BALASHOV, Z. G. 1962. *Ordovician nautiloids from the Siberian platform.* Izdatel'stvo Leningradskogo Universiteta, Leningrad, 205 pp. [In Russian].

—— 1975. Cephalopods of the Molodov and Kitaigorod formations of Podolia. *Voprosy Paleontologii*, **7**, 63–101 [In Russian].

—— and KISELEV, G. N. 1968. Some cephalopod molluscs from the Malinovetsky, Skal'sky, Borchovsky and Chortkovsky formations of Podolia. 7–29. *In* BALASHOV, Z. G. (ed.). *Silurian–Devonian fauna of Podolia*. Izdatel'stvo Leningradskogo Universiteta, Leningrad, 123 pp. [In Russian].

—— and ZHURAVLEVA, F. A. 1962. Order Orthoceratida. 82–93. *In* RUZHENCEV, V. E. (ed.). *Osnovy Paleontologii, Mollyuski-Golovonogie I*. Izdatel'stvo Akademiya Nauk SSSR, Moscow, 438 pp. [In Russian].

BANDEL, K. and STANLEY, G. D. 1989. Reconstruction and biostratinomy of Devonian cephalopods (Lamellorthoceratidae) with unique cameral deposits. *Senckenbergiana Lethaea*, **69**, 391–437.

BARRANDE, J. 1859. Über die organischen Ablagerungen in den Luft-Kammern der Orthoceraten. *Neues Jahrbuch für Mineralogie, Geognosie, Geologie und Petrefakten-Kunde*, **1859**, 780–798.

—— 1860. Troncature normale ou périodique de la coquille des certains céphalopodes paléozoiques. *Bulletin de la Societé Géologique de France, 2ème Série*, **17**, 573–601.

—— 1865–71. *Système Silurien du centre de la Bohême, I.ère partie: Recherches Paléontologiques, Volume II, Classe de Mollusques, Ordre des Céphalopodes*. Published privately, Prague, 1, 712 pp. (1867); 2, 266 pp. (1870); 3, 804 pp. (1874); 4, 742 pp., 5, 743 pp. (1877); 6, pls 1–107 (1865); 7, pls 108–244 (1866); 8, pls 245–350 (1868); 9, pls 351–460 (1870).

BARSKOV, I. S. 1960. Silurian and Devonian nautiloids from southern Fergana. *Byulleten Moskovskogo Obshchevstva Ispytatelei Prirody Otdel Geologicheskii*, **35** (4), 153–154. [In Russian]

—— 1963. System and phylogeny of the Pseudorthoceratids. *Byulleten Moskovskogo Obshchevstva Ispytatelei Prirody Otdel Geologicheskii*, **38** (4), 149–150. [In Russian].

—— 1966. *Cephalopods of the Late Ordovician and Silurian of Kazakhstan and Middle Asia*. Autoreferat dissertatsii na soiskanie utchenoyi stepeni kandidata geologicheskii mineralogicheskii nauka. Izdatel'stvo Moskovskogo Universitate, Moscow, 200 pp. [In Russian].

—— 1972. *Late Ordovician and Silurian cephalopod molluscs of Kazakhstan and Middle Asia*. Izdatel'stvo Akademiya Nauk SSR, Moscow, 109 pp. [In Russian].

—— and KISELEV, G. N. 1970. On the revision of some Silurian michelinoceratins (Cephalopoda, Orthocerida). *Paleontological Journal*, **1970** (3), 66–70.

BAYLE, E. and ZEILLER, R. 1878. *Fossiles principaux des terrains huiller. I. Explication de la carte géologique de la France, 4, atlas*. Imprimerie Nationale, Paris, pls 1–157.

BELKA, Z., KLUG, C., KAUFMANN, B., KORN, D., DÖRING, S., FEIST, R. and WENDT, J. 1999. Devonian conodont and ammonoid succession of the eastern Tafilalt (Quidane Chebbi section), Anti Atlas, Morocco. *Acta Geologica Polonica*, **49**, 1–23.

BENGTSON, P. 1988. Open nomenclature. *Palaeontology*, **31**, 223–227.

BEYRICH, H. 1850. *Arthrophyllum*. *Zeitschrift der Deutschen Geologischen Gesellschaft*, **2**, 10.

BILLINGS, E. 1857. New species of fossils from Silurian rocks of Canada. *Canada Geological Survey, Report of Progress 1853–1856. Report for the year 1856*, 247–345.

BLAKE, J. F. A. 1882. *Monograph of the British fossil Cephalopoda, Part 1, Introduction and Silurian species*. J. Van Voorst, London, iv + 248 pp., pls 1–31.

BLUMENBACH, J. F. 1803. *Specimen Archaeologiae Telluris Terra runque Inprimis Hannoveranum*. H. Dieterich, Göttingen, 28 pp., 3 pls.

BOGOSLOVSKY, B. I. 1969. Devonian ammonoids. I. Agoniatids. *Trudy Paleontologicheskogo Instituta, Akademiya Nauk SSSR*, **124**, 341 pp. [In Russian].

BRACHERT, T., BUGGISCH, W., FLÜGEL, E., HÜSSNER, H. M., JOACHIMSKI, M. M., TOURNEUR, F. and WALLISER, O. H. 1992. Controls of mud mound formation: the Early Devonian Kess-Kess carbonates of the Hamar Laghdad, Anti-Atlas, Morocco. *Geologische Rundschau*, **81**, 15–44.

BUGGISCH, W. and MANN, U. 2004. Carbon isotope stratigraphy of Lochkovian to Eifelian limestones from the Devonian of central and southern Europe. *International Journal of Earth Science*, **93**, 521–541.

BULTYNCK, P. and WALLISER, O. H. 2000. Devonian boundaries in the Moroccan Anti-Atlas. *Courier Forschungsinstitut Senckenberg*, **225**, 211–226.

CHANG ZHI-DUN 1962. Some species of nautiloids from the middle Ordovician of the Kuansyan region. *Acta Palaeontologica Sinica*, **10**, 514–523. [In Chinese, English summary].

CHAO, A. 1984. Non-parametric estimation of the number of classes in a population. *Scandinavian Journal of Statistics*, **11**, 265–270.

CHAPMAN, F. 1914. *Australian fossils*. G. Robertson and Co., Melbourne and Sydney, 341 pp.

CHAZDON, R. L., COLWELL, R. K., DENSLOW, J. S. and GUARIGUATA, M. R. 1998. Statistical methods for estimating species richness of woody regeneration in primary and secondary rain forests of NE Costa Rica. 285–309. *In* DALLMEIER, F. and COMISKEY, J. A. (eds). *Forest biodiversity research, monitoring and modeling: conceptual background and Old World case studies*. Parthenon Publishing, Paris, 671 pp.

CHEN JUN-YUAN 1975. The nautiloid fossils from the Qomolangma Feng region. 267–308. *In* TIBETAN SCIENTIFIC EXPEDITIONAL TEAM. *A report of scientific expedition in the Mount Jolmo Lungma, Palaeontology, 1*, Science Press, Beijing, 437 pp. [In Chinese].

CHLUPÁČ, I. 1985. Comments on the Lower–Middle Devonian boundary. *Courier Forschungsinstitut Senckenberg*, **75**, 389–400.

—— 1993. *Geology of the Barrandian – a field trip guide*. Verlag Waldemar Kramer, Frankfurt am Main, 163 pp.

—— 1999. Pragian/Zlíchovian and Zlíchovian/Dalejan boundary sections in the Lower Devonian of the Barrandian area, Czech Republic. *Newsletters on Stratigraphy*, **37**, 75–100.

—— and KUKAL, Z. 1996. Reflection of possible global Devonian events in the Barrandian area. 169–179. *In* WALLISER, O. H. (ed.). *Global events and event stratigraphy in the Phanerozoic*. Springer, Berlin, 333 pp.

——GALLE, A., HLADIL, J. and KALVODA, J. 2000. Series and stage boundaries in the Devonian of the Czech Republic. *Courier Forschungsinstitut Senckenberg*, **225**, 159–172.

CLARKE, K. R. and WARWICK, R. M. 1998. A taxonomic distinctness index and its statistical properties. *Journal of Applied Ecology*, **35**, 523–531.

——— 1999. The taxonomic distinctness measure of biodiversity: weighting of step lengths between hierarchical levels. *Marine Ecology Progress Series*, **184**, 21–29.

COLWELL, R. K. 2005. *EstimateS: Statistical estimation of species richness and shared species from samples. Version 7.5.* User's Guide and Application, published at: http://purl.oclc.org/estimates.

CONRAD, T. A. 1843. Observations on the lead bearing limestone of Wisconsin, and descriptions of a new genus of trilobites and fifteen new Silurian fossils. *Proceedings of the Academy of Natural Sciences of Philadelphia*, **1**, 329–335.

CUVIER, G. 1797–98. *Tableau élémentaire de l'histoire naturelle des animaux.* Badouin, Paris, 14 pp.

DZIK, J. 1984. Phylogeny of the Nautiloidea. *Palaeontologia Polonica*, **45**, 1–203.

—— and KORN, D. 1992. Devonian ancestors of *Nautilus*. *Paläontologische Zeitschrift*, **66**, 81–98.

ERBEN, H. K. 1960. Primitive ammonoidea aus dem Unterdevon Frankreichs und Deutschlands. *Neues Jahrbuch für Geologie und Paläontology, Abhandlungen*, **110**, 1–128.

—— 1964. Die Evolution der ältesten Ammonoidea (Lieferung I). *Neues Jahrbuch für Geologie und Paläontologie, Abhandlungen*, **120**, 107–212.

EVANS, D. H. 1994. Irish Silurian cephalopods. *Irish Journal of Earth Sciences*, **13**, 113–148.

FERRETI, A., GNOLI, M. and VAI, G. B. 1999. Silurian to Lower Devonian communities of Sardinia. 271–281. *In* BOUCOT, A. J. and LAWSON, J. D. (eds). *Paleocommunities – a case study from the Silurian and Lower Devonian.* Cambridge University Press, Cambridge, 859 pp.

FISCHER, P. 1880–87. *Manuel de Conchyliologie et de Paléontologie conchyliologique.* Savy, Paris, 896 pp.

FLOWER, R. H. 1936. Cherry valley cephalopods. *Bulletins of American Paleontology*, **22** (76), 273–372.

—— 1938. Devonian brevicones of New York and adjacent areas. *Palaeontographica Americana*, **2** (9), 1–84.

—— 1939. Study of Pseudorthoceratidae. *Palaeontographica Americana*, **11** (10), 214 pp.

—— 1940. Some Devonian Actinoceroidea. *Journal of Paleontology*, **14**, 442–442.

—— 1945. Classification of Devonian nautiloids. *American Midland Naturalist*, **33**, 675–724.

—— 1946. Ordovician cephalopods from the Cincinnati region. Part 1. *Bulletins of American Paleontology*, **29** (116), 85–751.

—— 1949. New genera of Devonian nautiloids. *Journal of Paleontology*, **23**, 74–80.

—— 1962. Notes on the Michelinoceratida. *New Mexico Institute of Mining and Technology, State Bureau of Mines and Mineral Resources, Memoir*, **10**, 21–40.

—— and CASTER, K. E. 1935. The cephalopod fauna of the Conewango Series of the Upper Devonian in New York and Pennsylvania. *Bulletins of American Paleontology*, **22** (75), 1–74.

—— and KUMMEL, B. 1950. A classification of the Nautiloidea. *Journal of Paleontology*, **24**, 604–616.

FOERSTE, A. F. 1925. Cephalopoda from Nesnayemi and Sulmeneva fjords in Novaya Zemlya collected by V. Roussanoff. [Report of the Scientific Results of the Norwegian Expedition to Novaya Zemlya 1921]. *Det Norske Videnskaps-Akademi i Oslo*, **31**, 1–38.

—— 1926. Actinosiphonate, trochoceroid and other cephalopods. *Denison University Bulletin, Journal of the Scientific Laboratories*, **21**, 285–383.

—— 1927. Devonian cephalopods from Alpena in Michigan. *Contributions from the Museum of Geology, University of Michigan*, **2** (9), 189–208.

—— 1928. A restudy of American orthoconic Silurian cephalopods. *Denison University Bulletin, Journal of the Scientific Laboratories*, **23**, 236–320.

—— 1929. Devonian cephalopods from the Moose River Basin. *Ontario Department of Mines, 37th Annual Report*, **37** (6), 70–78.

—— 1930. Port Byron and other Silurian cephalopods. *Denison University Bulletin, Journal of the Scientific Laboratories*, **23**, 1–110.

—— 1932. Black River and other cephalopods from Minnesota, Wisconsin, Michigan, and Ontario (Part 1). *Journal of the Scientific Laboratories of Denison University*, **27**, 47–136.

FRECH, F. 1897. *Lethaea geognostica. 1. Theil. Lethaea palaeozoica, 2.* Schweitzerbart'sche Verlagsbuchandlung, Stuttgart, 788 pp., 68 pls + 30 pls (subsequently numbered).

FOORD, A. H. 1888. *Catalogue of the fossil Cephalopoda in the British Museum (Natural History). Part 1.* British Museum, London, 344 pp.

GABBOTT, S. E. 1999. Orthoconic cephalopods and associated fauna from the Late Ordovician Soom Shale Lagerstätte, South Africa. *Palaeontology*, **42**, 123–148.

GARCIA-ALCALDE, J., TRUYÓLS-MASSONI, M., PARDO-ALONSO, M., BULTYNCK, P. and CARLS, P. 2000. Devonian chronostratigraphy of Spain. *Courier Forschungsinstitut Senckenberg*, **225**, 131–144.

GAUDRY, A. 1883. *Les enchaînements du monde animal de les temps géologiques, fossiles primaires.* Savy, Paris, 317 pp.

GNOLI, M. 1982. Lower Devonian orthocone cephalopods from Iglesiente and Sulcis regions (southwestern Sardinia). *Bollettino della Società Paleontologica Italiana*, **21**, 73–98.

—— 1990. New evidence for faunal links between Sardinia and Bohemia in Silurian time on the basis of nautiloids. *Bollettino della Società Paleontologica Italiana*, **29**, 289–307.

—— 1998. Some remarks and emendation of the family Arionoceratidae Dzik, 1984 (Cephalopoda, Nautiloidea). *Palaeontologia Electronica*, **1** (2), 7. http://palaeo-electronica.org/1998_2/gnoli/issue2.htm.

—— 2003. Northern Gondwanan Siluro–Devonian palaeogeography assessed by cephalopods. *Palaeontologia Electronica*, **5** (2), 19. http://palaeo-electronica.org/paleo/2002_2/gondwana/issue2_02.htm.

—— and KISELEV, G. 1994. Revision of the family Sphooceratidae Flower, 1962 (Orthocerida). *Bolletino della Società Paleontologica Italiana*, **33**, 415–420.

—— and SERPAGLI, E. 1977. Silurian cephalopods of the Meneghini collection (1857) with reproduction of the original plates. *Bollettino della Società Paleontologica Italiana*, **30**, 492–503.

——— 1991. Nautiloid assemblages from middle–late Silurian of southwestern Sardinia: a proposal. *Bollettino della Società Paleontologica Italiana*, **30**, 187–195.

—— PAREA, G. C., RUSSO, F. and SERPAGLI, E. 1979. Paleoecological remarks on the 'Orthoceras limestone' of southwestern Sardinia (middle Upper Silurian). *Memorie della Società Geologica Italiana*, **20**, 405–423.

GÓMEZ-ALBA, J. A. S. 1988. *Guía de Campo de los Fósiles de Espana y de Europa*. Ediciones Omega, S.A., Barcelona, 972 pp.

GORDON, M. 1960. Some American Midcontinent Carboniferous cephalopods. *Journal of Paleontology*, **34**, 133–151.

GRABAU, A. W. and SHIMER, H. W. 1910. *North American index fossils, Invertebrates, 2*. Seiler et Cie, New York, NY, 909 pp.

HALL, J. 1877 (imprint 1876). Illustrations of Devonian fossils: Gastropoda, Pteropoda, Cephalopoda, Crustacea and corals of the Helderberg, Hamilton and Chemung groups. *New York Geological Survey, Palaeontology of New York*, **3**, 7 pp., 133 pls.

—— 1879. *Natural History of New York, Paleontology, Volume 5, Part 2, containing descriptions of the Gastropoda, Pteropoda, Cephalopoda, Crustacea and corals of the Upper Helderberg, Hamilton Portage and Chemung groups*. New York Geological Survey, Albany, 492 pp.

HAUDE, R. and WALLISER, O. H. 1998. Conodont based Upper Silurian–Lower Devonian range of scyphocrinoids in SE Morocco. *Temas Geológico-Mineros ITGE*, **23**, 94–96.

HELLER, T. 1925. Die Fauna des Obersilurischen Orthocerenkalks von Elbersreuth. *Geognostische Jahreshefte*, **38**, 1–197.

HERITSCH, F. 1931. Die obersilurische Fauna des Wiedatales im Harz. *Jahrbuch der Preussischen Geologischen Landesanstalt*, **50**, 514–580.

HEWITT, R. A. and WATKINS, R. 1980. Cephalopod ecology across a late Silurian shelf tract. *Neues Jahrbuch für Geologie und Paläontologie, Abhandlungen*, **160**, 96–117.

HISTON, K. 1998. Die Nautiloideen-Fauna aus dem Silur der Karnischen Alpen. *Geologisch-Paläontologische Mitteilungen*, **23**, 105–115.

—— 1999. Revision of Silurian nautiloid cephalopods from the Carnic Alps. – The Heritsch (1929) collection in the Geological Survey of Austria. *Abhandlungen der Geologischen Bundesanstalt*, **56**, 229–259.

—— and GNOLI, M. 1999. Nautiloid paleobathymetry from the Silurian "Orthoceras Limestone" facies of SW Sardinia. 381–394. *In* OLÓRIZ, F. and RODRIGUEZ-TOVAR, F. J. (eds). *Advancing research on living and fossil cephalopods*. Kluwer Academic/Plenum Publishers, New York, NY, 535 pp.

HLADÍKOVÁ, J., HLADIL, J. and KŘÍBEK, B. 1997. Carbon and oxygen isotope record across the Přídolí to Givetian stage boundaries in the Barrandian Basin (Czech Republic). *Palaeogeography, Palaeoclimatolology, Palaeoecology*, **132**, 225–241.

HOLLAND, C. H. 2000. *Harrisoceras* and *Temperoceras*, closely related cephalopod genera from the British Silurian. *Bollettino della Società Paleontologica Italiana*, **39**, 117–122.

——2003. Some observations on bactritid cephalopods. *Bulletin of Geoscience*, **78**, 369–372.

—— GNOLI, M. and HISTON, K. 1994. Concentrations of Palaeozoic nautiloid cephalopods. *Bolletino della Società Paleontologica Italiana*, **33**, 83–99.

HOLLARD, H. 1977. Le domaine de l'Anti-Atlas au Maroc. 168–194. *In* MARTINSSON, A. (ed.). *The Silurian-Devonian boundary. Final report of the committee on the Silurian-Devonian boundary within IUGS Commission on Stratigraphy and a state of the art report for Project Ecostratigraphy*. IUGS Series A, **5**, 1–394.

—— 1981. Tableux de corrélation du Silurien et Devonien de l'Anti-Atlas. *Notes du Service Géologique du Maroc*, **42** (308), 23–46.

HOLZAPFEL, E. 1885. Das obere Mitteldevon (Schichten mit *Stringocephalus burtini* und *Maenioceras terebrum*) im Rheinschen Gebirge. *Königlich Preussische Landesanstalt, Abhandlungen, Neue Folge*, **16**, 1–460.

HYATT, A. 1883–84. Genera of fossil cephalopods. *Proceedings of the Boston Society of Natural History*, **22**, 273–338.

—— 1900. Cephalopoda. 502–592. *In* ZITTEL, K. A. VON. *Textbook of Paleontology, 1*. Second edition, translated and edited by Eastmann, C. R. Macmillan and Co., London, 839 pp.

JOHNSON, J. G., KLAPPER, G. and SANDBERG, C. A. 1985. Devonian eustatic fluctuations in Euramerica. *Geological Society of America, Bulletin*, **96**, 567–587.

KAYSER, E. 1878. Die Fauna der ältesten Devon-Ablagerungen des Harzes. *Abhandlungen zur Geologischen Specialkarte von Preussen und den Thüringischen Staaten*, **2** (4), 1–295.

KISELEV, G. N. 1969. *Silurian cephalopoda of the Bolshezemelskaya Tundra and the north of the Urals*. Autoreferat dissertatsii na soiskanie utchenoyi stepeni kandidata geologicheskii mineralogicheskii nauka Leningradskii Gosudarstvenyi Universitet, University of Leningrad/St. Petersburg, 22 pp. [In Russian].

—— 1971. Embryonic shells of the Silurian Michelinoceratidae. *Voprosy Paleontologii*, **6**, 41–51. [In Russian].

—— 1972. Review on 'Revision of the Orthoceratids' (Zur Revision der Orthoceratidae by H. Ristedt). *Paleontological Journal*, **6** (5), 427–428.

—— 1973. New data on the structure of the connecting rings in Silurian michelinoceratins (Orthocerida). *Paleontologicheskiy Zhurnal*, **1973** (2), 124–126. [In Russian]

—— 1985. Discoveries of Palaeozoic orthoceratids with preserved first chambers in Novaya Zemlya and Siberia. 78–88. *In* BONDAREV, V. I. (ed.). *Stratigraphy and fauna of the Palaeozoic of Novaya Zemlya*. Ministerstvo Geologii PGO 'SEVMORGEOLOGIA', Leningrad/St. Petersburg, 121 pp. [In Russian].

—— 1991. On ontogenesis and phylogenesis of orthoconic cephalopods. *Vestnik Leningradskogo Universiteta, Seriya 7, Geologiya Geografiya*, **1991** (1), 18–27. [In Russian].

—— and EVSEEV, K. P. 1982. Late Silurian orthocones from the 'Reki Kari' Basin (Polar Urals) and their stratigraphical significance. *Vestnik Leningradskogo Universiteta*, **1982** (12), 110–113. [In Russian].

—— MIRONOYA, M. G. and SINITSINA, I. N. 1987. *Atlas of the Silurian molluscs of Podolia*. Izdatel'stvo Leningradskogo Universiteta, Leningrad/St. Petersburg, 180 pp. [In Russian].

KLUG, C. 2001. Early Emsian ammonoids from the eastern Anti-Atlas (Morocco) and their succession. *Paläontologische Zeitschrift*, **74**, 479–515.

—— 2002. Quantitative stratigraphy and taxonomy of late Emsian and Eifelian ammonoids of the eastern Anti-Atlas (Morocco). *Courier Forschungsinstitut Senckenberg*, **238**, 1–109.

—— KRÖGER, B., KORN, D., RÜCKLIN, M., SCHEMM-GREGORY, M., DE BAETS, K. and MAPES, R. 2008. Ecological change during the early early Emsian (Devonian) in the Tafilalt (Morocco), the origin of the Ammonoidea, and the first African pyrgocystid edrioasteroids, machaerids and phyllocarids. *Palaeontographica, Abteilung A*, **283**, 1–94.

KOLEBABA, I. 1975. *Caliceras* n. gen. and ontogeny of *C. capillosum* (Barrande) (Nautiloidea, Michelinoceratida). *Časopis pro Mineralogii a Geologii*, **20**, 377–391.

—— 1977. New information on the longitudinally sculptured orthoceroids. *Časopis pro Mineralogii a Geologii*, **22**, 125–138.

—— 1999. Sipho-cameral structures in some Silurian cephalopods from the Barrandian area. *Sborník Národního Muzea v Praze Řada B, Přírodní Vedy*, **55**, 1–15.

—— 2002. A contribution to the theory of the cameral mantle in some Silurian Nautiloidea (Mollusca, Cephalopoda). *Vestnik Českého Geologického Ústavu*, **77** (3), 183–186.

KORN, D. and KLUG, C. 2003. Morphological pathways in the evolution of Early and Middle Devonian ammonoids. *Paleobiology*, **29**, 329–348.

KŘÍŽ, J. 1998. Recurrent Silurian–lowest Devonian cephalopod limestones of Gondwanan Europe and Perunica. 183–198. *In* LANDING, E. and JOHNSON, M. (eds). *Silurian cycles. Linkages of dynamic stratigraphy with atmospheric, oceanic, and tectonic changes*. New York State Museum, Bulletin, **491**, 183–199.

—— 1999. Bivalvia dominated communities of Bohemian type from the Silurian and Lower Devonian carbonate facies. 229–253. *In* BOUCOT, A. J. and LAWSON, J. D. (eds). *Paleocommunities – a case study from the Silurian and Lower Devonian*. Cambridge University Press, Cambridge, 859 pp.

—— and FERRETTI, A 1995. Cephalopod limestone biofacies in the Silurian of the Prague Basin, Bohemia. *Palaios*, **10**, 249–253.

KRÖGER, B. 2003. The size of the siphuncle in cephalopod evolution. *Senckenbergiana Lethaea*, **83**, 39–52.

—— 2004. Revision of Middle Ordovician orthoceratacean nautiloids from Baltoscandia. *Acta Palaeontologica Polonica*, **49**, 57–74.

—— and ISAKAR, M. 2006. Revision of annulated orthoceridan cephalopods of the Baltoscandic Ordovician. *Fossil Record*, **9**, 137–163.

—— and MAPES, R. H. 2007. On the origin of bactritoids. *Paläontologische Zeitschrift*, **81**, 316–327.

—— BERESI, M. and LANDING, E. 2007. Early orthoceratoid cephalopods from the Argentine Precordillera (Lower–Middle Ordovician). *Journal of Paleontology* **82**, 1263–1280.

—— KLUG, C. and MAPES, R. 2005. Soft-tissue attachments in orthocerid and bactritid cephalopods from the Early and Middle Devonian of Germany and Morocco. *Acta Palaeontologica Polonica*, **50**, 329–342.

KUHN, O. 1940. *Paläozoologie in Tabellen*. Fischer Verlag, Jena, 50 pp.

KUMMEL, B., FURNISH, W. M. and GLENISTER, B. F. 1964. Nautiloidea-Nautilida. K383–K457. *In* MOORE, R. C. (ed.). *Treatise on invertebrate paleontology, Part K, Mollusca 3, Cephalopoda*. Geological Society of America, Boulder, CO, and University of Kansas Press, Lawrence, KS, xxvii + 519 pp.

LAI CAI-GEN 1960. Nautiloid fossils of the *Yangtzeella poloi* beds from Shensi and Hupei. *Acta Palaeontologica Sinica*, **8**, 251–271. [In Chinese, English summary].

—— 1986. On the Lituitidae. *Bulletin of the Chinese Academy of Geological Siences*, **12**, 107–126. Beijing. [In Chinese, English summary]

LEGRAND, P. 2003. Silurian stratigraphy and paleogeography of the northern African margin of Gondwana. 59–104. *In* LANDING, E. and JOHNSON, M. E. (eds). *Silurian lands and seas. Paleogeography outside of Laurentia*. New York State Museum, Bulletin, **493**, 400 pp.

MAGURRAN, A. E. 2004. *Measuring biological diversity*. Blackwell Science, Oxford, 260 pp.

MANDA, S. 2001. Some new or little known cephalopods from the Lower Devonian Pragian carbonate shelf (Prague Basin, Bohemia) with remarks on Lochkovian and Pragian cephalopod evolution. *Journal of the Czech Geological Society*, **46**, 269–286.

MAREK, J. 1998. Pallioceratida ordo n. – a new order of the Palaeozoic cephalopods (Mollusca, Cephalopoda). *Vestnik Českého Geologického Ústavu*, **73** (2), 181–182.

MASCKE, H. 1876. *Clinoceras*, n. g., ein silurischer Nautilide mit gelappten Scheidewänden. *Zeitschrift der Deutschen Geologischen Gesellschaft*, **28**, 49–55.

MAURER, F. 1876. Paläontologische Studien im Gebiet des rheinischen Devons. 3. Die Thonschiefer des Ruppbachthales bei Diez. *Neues Jahrbuch für Mineralogie, Geologie und Paläontologie*, **1876**, 808–848.

—— 1881. Paläontologische Studien im Gebiet des rheinischen Devons. *Neues Jahrbuch für Mineralogie, Geologie und Paläontologie*, **1881** (1), 1–112.

M^C COY, F. 1844. *A synopsis of the characters of the Carboniferous Limestone fossils of Ireland*. University Press, Dublin, 274 pp, 29 pls.

MEHL, J. 1984. Radula and arms of *Michelinoceras* sp. from the Silurian of Bolivia. *Paläontologische Zeitschrift*, **58**, 211–229.

MENEGHINI, G. 1857. Paléontologie de l'Ile de Sardaigne. 53–144, pl. C. *In* LA MARMORA, A. (ed.). *Voyage en Sardaigne*. Imprimérie Royal, Turin, 584 pp.

MILLER, A. K. 1931. Two new genera of Late Paleozoic cephalopods from Central Asia. *American Journal of Science*, **22**, 417–425.

MILLER, S. A. 1877. *The American Palaeozoic fossils: a catalogue of the genera and species with names of authors, dates, places of publication, groups of rocks in which found, and the etymology and signification of the words and an introduction devoted to the stratigraphical geology of the Palaeozoic rocks.* Published privately, Cincinnati, OH, 179 pp.

NIKO, S. 1996. Pseudorthoceratid cephalopods from the Early Devonian Fukuji Formation of Gifu Prefecture, Central Japan. *Transactions and Proceedings of the Palaeontological Society of Japan, New Series*, **181**, 347–360.

PORTLOCK, J. E. 1843. *Report on the geology of the county of Londonderry and parts of Tyrone and Fermanagh*. Milliken, Dublin, 784 pp.

QUENSTEDT, F. A. 1845–49. *Petrefactenkunde Deutschlands. 1. Abteilung, 1. Band, Cephalopoden*. Fues Verlag, Tübingen, 580 pp.

ROEMER, C. F. 1844. *Das rheinische Uebergangsgebirge. Eine palaeontologische-geognostische Darstellung*. Verlag der Hahn'schen Hofbuchhandlung, Hannover, 96 pp.

RISTEDT, H. 1968. Zur Revision der Orthoceratidae. *Akademie der Wissenschaften und Literatur in Mainz, Abhandlungen der Mathematisch-Naturwissenschaftlichen Klasse*, **1968**, 211–287.

—— 1981. Bactriten aus dem Obersilur Böhmens. *Mitteilungen des Geologisch-Paläontologischen Instituts der Universität Hamburg*, **51**, 23–26.

ROBARDET, M. and GUTIÉRREZ-MARCO, J. C. 2004. The Ordovician, Silurian and Devonian sedimentary rocks of the Ossa-Morena Zone (SW Iberian Peninsula, Spain). *Journal of Iberian Geology*, **30**, 73–92.

SAEMANN, L. 1852. Über die Nautiliden. *Palaeontographica*, **3**, 121–167.

SANDBERGER, G. 1843. Allgemeine Schilderung der paläontologischen Verhältnisse der älteren Formationen Nassaus. *Amtlicher Bericht über die Versammlung der Deutschen Naturforscher und Ärzte zu Mainz*, **20**, 154–160.

—— and SANDBERGER, F. 1849–56. *Die Versteinerungen des rheinischen Schichtensystems in Nassau*. Kreidel und Niedner Verlagsbuchhandlung, Wiesbaden, 564 pp., 41 pls.

SCHINDEWOLF, O. H. 1932. Zur Stammesgeschichte der Ammoneen. *Paläontologische Zeitschrift*, **14**, 164–181.

—— 1933. Vergleichende Morphologie und Phylogenie der Anfangskammern tetrabranchiater Cephalopoden – Eine Studie über Herkunft, Stammesentwicklung und System der niederen Ammonoideen. *Abhandlungen der Preussischen Geologischen Landesanstalt, Neue Folge*, **148**, 1–115.

SCHÖNLAUB, H. P. 1996. Significant geological events in the Paleozoic record of the Southern Alps (Austrian part). 163–168. *In* WALLISER, O. H. (ed.). *Global events and event stratigraphy in the Phanerozoic*. Springer, Berlin, 333 pp.

SERPAGLI, E. and GNOLI, M. 1977. Upper Silurian cephalopods from southwestern Sardiania. *Bollettino della Società Paleontologica Italiana*, **16**, 153–196.

SHIMANSKY, V. N. 1951. On the question of the extinction of Upper Palaeozoic cephalopods. *Doklady Akademiya Nauk SSSR*, **79**, 867–870. [In Russian].

—— 1962. Order Nautilida. 115–154. *In* RUZHENCEV, V. E. (ed.). *Osnovy Paleontologii, Mollyuski-Golovonogie I*. Izdatelstvo Akademiya Nauk SSR, Moscow, 438 pp. [In Russian].

SHIMIZU, S. and OBATA, T. 1935a. New genera of Gotlandian and Ordovician nautiloids. *Journal of the Shanghai Science Institute, Section 2*, **2**, 1–10.

——1935b. On a new Ordovician nautiloid genus *Sinoceras. Proceedings of the Imperial Academy, Tokyo*, **11**, 324–325.

SOWERBY, J. DE C. 1839. Fossil shells of the upper Ludlow rock. 608–644. *In* MURCHISON, R. I. (ed.). *The Silurian System*. John Murray, London, xx + 786 pp.

STANLEY, G. D. and TEICHERT, C. 1976. Lamellorthoceratids (Cephalopoda, Orthoceratoidea) from the Lower Devonian of New York. *University of Kansas, Paleontological Contributions, Paper*, **86**, 1–14.

STAROBOGATOV, Y. I. 1974. Xenoconchias and their bearing on the phylogeny and systematics of some molluscan classes. *Paleontological Journal*, **8**, 1–13.

STEININGER, J. 1853. *Geognostische Beschreibung der Eifel*. Lintz'sche Buchhandlung, Trier, 143 pp.

STOKES, C. 1838. On some species of Orthocerata. *London and Edinburgh Philosophical Magazine*, **13**, 388–390.

—— 1840. On some species of Orthocerata. *Transactions of the Geological Society of London, Series 2*, **5**, 705–714.

SWEET, W. C. 1964a. Nautiloidea-Oncocerida. K277–K319. *In* MOORE, R. C. (ed.). *Treatise on invertebrate paleontology, Part K, Mollusca 3, Cephalopoda*. Geological Society of America, Boulder, CO, and University of Kansas Press, Lawrence, KS, xxvii + 519 pp.

—— 1964b. Nautiloidea-Orthocerida. K216–K261. *In* MOORE, R. C. (ed.), *Treatise on invertebrate paleontology, Part K, Mollusca 3, Cephalopoda*. Geological Society of America, Boulder, CO, and University of Kansas Press, Lawrence, KS, xxvii + 519 pp.

TALENT, J. A., MAWSON, R., ANDREW, A. S., HAMILTON, J. and WHITFORD, D. J. 1993. Middle Paleozoic extinction events: faunal and isotopic data. *Palaeogeography, Palaeoclimatology, Palaeoecology*, **103**, 139–152.

TEICHERT, C. 1933. Der Bau der actinoceroiden Cephalopoden. *Palaeontographica, Abteilung A*, **77**, 111–230.

——1939. Nautiloid cephalopods from the Devonian of Western Australia. *Journal of the Royal Society of West Australia*, **25**, 103–121.

—— 1961. Les Nautiloides des genres *Arthrophyllum* Beyrich et *Lamellorthoceras* Term. and Term. *Annales de Paléontologie*, **47**, 91–113.

—— 1964. Actinoceratoidea. K190–K216. *In* MOORE, R. C. (ed.). *Treatise on invertebrate paleontology, Part K, Mollusca 3, Cephalopoda*. Geological Society of America, Boulder, CO, and University of Kansas Press, Lawrence, KS, xxvii + 519 pp.

—— and GLENISTER, B. F. 1952. Fossil nautiloid faunas from Australia. *Journal of Paleontology*, **26**, 730–752.

TERMIER, H. and TERMIER, G. 1950. Paléontologie Marocaine. II. Invértèbres de l'ère primaire. Fasc. III, Mollusques. *Notes et Mémoires du Service Carte Géologique du Maroc*, **78**, 246 pp.

TSHERNYSHEV, T. 1885. Der permische Kalkstein im Gouvernement Kostroma. *Verhandlungen der Mineralischen Gesellschaft St. Petersburg, 2 Serie*, **20**, 265–315.

TSOU SI-PING 1966. Middle Silurian nautiloids from Guang-yuan, Szechuan Province. *Acta Palaeontologica Sinica*, **14**, 10–37. [In Chinese, English summary].

TUREK, V. and MAREK, J. 1986. Notes on the phylogeny of the Nautiloidea. *Paläontologische Zeitschrift*, **60**, 245–253.

VANUXEM, L. 1842. *Geology of New York, Part III, comprising the survey of the third geological district, Albany*. C. Van Benthuysen, Albany, NY, 306 pp.

VERNEUIL, E. P. DE 1850. Idées générales sur l'ensemble du terrain paléozoique de la Sarthe, avec une liste de fossiles. *Bulletin de la Societé Géologique de France, 2ème Série*, **7**, 769–787.

VIDAL, L. M. 1914. Nota paleontológica sobre el Silúrico superior del Pirineo catalán. *Memorias de la Real Academia de Ciencias y Artes de Barcelona, 3*, **11** (19), 307–313.

WALLISER, O. H. 1985. Natural boundaries and commission boundaries in the Devonian. *Courier Forschungsinsititut Senckenberg*, **75**, 401–408.

WENDT, J., AIGNER, T. and NEUGEBAUER, J. 1984. Cephalopod limestone deposition on a shallow pelagic ridge: the Tafilalt Platform (upper Devonian, eastern Anti-Atlas, Morocco). *Sedimentology*, **31**, 601–625.

WHIDBORNE, G. F. 1890. A monograph of the Devonian fauna of the south of England. Volume 1, part 2. *Monographs of the Palaeontographical Society*, 47–154, pls 5–15.

WILMSEN, M. 2006. Origin and significance of Upper Cretaceous bioevents: examples from the Cenomanian. *Berichte des Instituts für Geowissenschaften Christian-Albrechts-Universität*, **22**, 151–152.

WILSON, A. E. 1961. Cephalopoda of the Ottawa Formation of the Ottawa-St. Lawrence Lowland. *Bulletin of the Geological Survey of Canada*, **67**, 1–106.

YÜ, C. C. 1930. The Ordovician Cephalopoda of Central China. *Palaeontologica Sinica, Series B*, **1** (2), 1–101.

ZHURAVLEVA, F. A. 1959. On the family Michelinoceratidae. *Materialii k Osnovam Paleontologii*, **3**, 47–48. [In Russian].

—— 1974. Devonian nautiloids. Orders Oncoceratida, Tarphyceratida, Nautilida. *Trudy Paleontologicheskogo Instituta, Akademiya Nauk SSSR*, **142**, 142 pp. [In Russian].

——1978. Devonian Orthoceratoidea. *Trudy Paleontologicheskogo Instituta, Akademiya Nauk SSSR*, **168**, 224 pp. [In Russian].

—— 1979. New Middle Devonian nautiloids from southern Transcaucasia and the Altai. *Paleontologicheskii Zhurnal*, **13** (3), 304–311. [In Russian].

—— 2000. New Early Devonian oncocerids (Cephalopoda) of Novaya Zemlya. *Paleontologicheskii Zhurnal*, **2000** (6), 18–25. [In Russian].

APPENDIX

Complete list of occurrences of non-ammonoid cephalopods from the Filon Douze section, Tafilalt, Morocco

Species	OP/ OP-M	OSP I-II*	PK	OJ-I	BS	OJ-II	CK*	KMO-I	KMO-II	KMO-III	EF	Deiroc. Lst.	KMO-IV*	Erben. Lst.	KMO-V*	F I+II	P I+II
Pseudendoplectoceras lahcani gen. et sp. nov.									10								
Brevicoceras magnum gen. et sp. nov.																	1
Cerovoceras fatimi gen. et sp. nov.																	1
Cerovoceras brevidomus gen. et sp. nov.																	1
Jovellania cheirae sp. nov.				8		1											
Bohemojovellania bouskai Manda, 2001				2													
Bohemojovellania adrae sp. nov.				5													
Bohemojovellania obliquum sp. nov.				3													
Ankyloceras sp.		1															
Ventrobalashovia zhuravlevai gen. et sp. nov.										1							
Mutoblakeoceras inconstans gen. et sp. nov.																1	
Taflaltoceras adgoi gen. et sp. nov.								1									
Indet Nothoceratidae sp. A																	1
Orthorizoceras desertum gen. et sp. nov.		1															
Centroceras sp.													1				
Ormoceras sp								1									
Deiroceras hollardi sp. nov.												5					
Metarmenoceras fatimae sp. nov.														3			
Getdoloceras ouaoufilalense gen. et sp. nov.									1	1							
Neocycloceras? termierorum sp. nov.													1			2	
Probatoceras? sp.									1								
Subdoloceras tafilaltense gen. et sp. nov.													1				

Appendix Continued.

Species	OP/ OP-M	OSP I-II*	PK	OJ-I	BS	OJ-II	CK*	KMO-I	KMO-II	KMO-III	EF	Deiroc. Lst.	KMO-IV*	Erben. Lst.	KMO-V*	F I+II	P I+II
Subdoloceras atrouzense gen. et sp. nov.									3	1							
Subdoloceras engeseri gen. et sp. nov.								1	15								
Subormoceras erfoudense gen. et sp. nov.														2		3	
Subormoceras rissaniense gen. et sp. nov.																1	1
Subormoceras sp.		1															
Spyroceras cyrtopatronus sp. nov.										2							
Spyroceras fukuijense? Niko, 1996										2							
Spyroceras latepatronus sp. nov.								3									
Spyroceras patronus (Barrande, 1866)								1	1	1							
Spyroceras? sp. A										1							
Spyroceras? sp. B										1							
Cancellspyroceras loricatum (Barrande, 1868)										2							
Diagoceras sp.																	
Suloceras longipulchrum sp. nov.										3							1
Arthrophyllum vermiculare (Termier and Termier, 1950)				15				37	58	14	6		4	4	1	1	
Sphooceras truncatum Barrande, 1860	5																
Chebbioceras erfoudense Klug et al., 2008											10						
Infundibuloceras brevinira Klug et al., 2008											6		1				
Infundibuloceras longicameratum sp. nov.													1				
Infundibuloceras mohamadi sp. nov.														4			1
Kopaninoceras? jucundum (Barrande, 1870)				2													

Taxon														
Kopaninoceras? dorsatoides sp. nov	78													
Kopaninoceras thyrsus (Barrande, 1870)	80	2												
Kopaninoceras? sp.	3													
Merocycloceras sp. A			1											
Merocycloceras sp. B							4							
Merocycloceras? sp. C			1											
Michelinoceras sp.	2													
Orthocycloceras? fluminense (Meneghini 1857)	4	4		1										
Orthocycloceras tafilaltense sp. nov.								3		11				
Pseudospyroceras reticulum gen. et sp. nov.		1						1						
Theoceras filondouzense gen. et sp. nov.														3
Tibichoanoceras tibichoanum gen. et sp. nov.							1							
Orthoceratidae gen. et sp. indet.												1		
Arionoceras canonicum (Meneghini, 1857)	188													
Arionoceras aff. *canonicum* (Meneghini, 1857)	65													
Arionoceras capillosum (Barrande, 1867)				25										
Arionoceras sp. A	7													
Adiagoceras taouzense gen. et sp. nov.			177			1								
Parakionoceras originale (Barrande, 1868)		4	2			1								
Anaspyroceras sp.														
Indet. Dawsonoceratidae									1					
Angeisonoceras reteornatum gen. et sp. nov.			6				1	5						
Temperoceras? ludense (Sowerby, *in* Murchison 1839)	6													
Temperoceras aequinudum sp. nov.					2		10	27	9	38	1			
Temperoceras migrans (Barrande, 1868)							24	2	3		1		3	

Appendix Continued.

Species	OP/ OP-M	OSP I-II*	PK	OJ-I	BS	OJ-II	CK*	KMO -I	KMO -II	KMO -III	EF	Deiroc. Lst.	KMO -IV*	Erben. Lst.	KMO -V*	F I+II	P I+II
Kionoceras? doricum (Barrande, 1868)	4																
Sichuanoceras zizense sp. nov.				2													
Sphaerorthoceras sp. A					1												
Sphaerorthoceras sp. B					1												
Akrosphaerorthoceras sp.		1			3												
Hemicosmorthoceras aichae sp. nov.				1													
Hemicosmorthoceras semimbricatum Gnoli, 1982	1	2	2	80			4										
Parasphaerorthoceras sp. B sensu Ristedt 1968					1												
Parasphaerorthoceras sp. G sensu Ristedt 1968			1														
Parasphaerorthoceras sp. H sensu Ristedt 1968	1				3												
Plagiostomoceras pleurotomum? (Barrande, 1866)	1																
Plagiostomoceras bifrons (Barrande, 1866)				2													
Plagiostomoceras culter (Barrande, 1866)	1		100														
Plagiostomoceras lategruenwaldti sp. nov.				2													
Plagiostomoceras reticulatum sp. nov.				9													
Bactrites gracile (Blumenbach, 1803)																3	
Bactrites? sp.										1							
Devonobactrites obliqueseptatus (Sandberger and Sandberger, 1852)											486						
Devonobactrites emsiense sp. nov.														3	1		
Devonobactrites sp.											3						
Lobobactrites ellipticus (Frech, 1897)																1	
Total number of specimens	442	16	103	318	36	3	6	84	127	42	560	5	10	16	6	15	7